CLONING WILD LIFE

BIOPOLITICS: MEDICINE, TECHNOSCIENCE,
AND HEALTH IN THE 21ST CENTURY
General Editors: Monica J. Casper and Lisa Jean Moore

Cloning Wild Life

Zoos, Captivity, and the Future of Endangered Animals

Carrie Friese

NEW YORK UNIVERSITY PRESS
New York and London

NEW YORK UNIVERSITY PRESS
New York and London
www.nyupress.org

References to Internet websites (URLs) were accurate at the time of writing.
Neither the author nor New York University Press is responsible for URLs that
may have expired or changed since the manuscript was prepared.

Library of Congress Cataloging-in-Publication Data

Friese, Carrie.
Cloning wild life : zoos, captivity, and the future of endangered animals / Carrie Friese.
pages cm
Includes bibliographical references and index.
ISBN 978-0-8147-2908-3 (cl : alk. paper) — ISBN 978-1-4798-3638-3 (pb : alk. paper)
1. Cloning. 2. Endangered species. I. Title.
QH442.2.F75   2013
571.9'646—dc23   2013009420

New York University Press books are printed on acid-free paper,
and their binding materials are chosen for strength and durability.
We strive to use environmentally responsible suppliers and materials
to the greatest extent possible in publishing our books.

Manufactured in the United States of America
10 9 8 7 6 5 4 3 2 1

For Stephanie . . . and our two cats and two dogs

CONTENTS

ACKNOWLEDGMENTS

First and foremost, I would like to thank all the people who took time out of their busy schedules to participate in this research. I have continually been inspired by the kindness and generosity of those I met while conducting this study. Cloning endangered animals has certainly been a contentious topic in zoos, but the people involved in these debates are extremely gracious. I consider myself lucky to have met so many interesting and inspiring people in the course of my work.

This project simply would not have been possible without guidance from a number of tremendous mentors. Adele Clarke's intellectual generosity and curiosity along with her sustained encouragement made it possible for this medical sociologist to study endeavors to clone endangered animals. Charis Thompson's work on assisted reproduction in the zoo inspired this research, and her acuity in interpreting the complexities and significances of these practices has never ceased to amaze—and help—me. Stefan Timmermans taught me to be a far more methodical thinker and writer, and helped me gain confidence in having decided to study such a seemingly obscure set of social practices as a sociologist. I would also like to acknowledge the special role of Gay Becker, who not only introduced me to the assisted reproductive technologies but also mentored me through much of the original research. She sadly passed away before I had completed the dissertation upon which this book is based. Janet Shim kindly stepped in to see me through, and has continued to help me find my way around the biopolitics of zoological parks. I consider myself incredibly lucky to have worked with and learned from so many generous scholars.

I was incredibly fortunate to have written this book while participating in a writing group with a number of exceptionally smart and inspiring women at the London School of Economics and Political

Science (LSE). Suki Ali, Sarah Franklin, Emily Jackson, and Ilina Singh generously read and commented on most of this book, providing crucial insights, helpful writing tips, and many laughs along the way. Your friendship made this book fun to write. A special thank you goes to Sarah Franklin, who both organized this group and has been a sustained source of inspiration and support.

Many thanks go to NYU Press. In particular, I would like to thank Ilene Kalish, Caelyn Cobb, Monica Casper, and Lisa Jean Moore. I would also like to thank the four anonymous reviewers whose comments were tremendously helpful in improving the book.

Over the years, a number of people have read and commented on material that ultimately found its way into this book. Special thanks go to Rene Almeling, Valentina Amorese, Olivia Banner, Soraya de Chadarevian, Alasdair Cochrane, Cassandra Crawford, Gail Dobel, Joe Dumit, Des Fitzgerald, Laura Foster, Chris Ganchoff, Nick Hopwood, Uffe Jensen, Brett Kious, Lene Koch, Chris Kortright, Ilana Löwy, Michael Lynch, Ed McCabe, Emily Martin, Aaron Panofsky, Suzanne Pelka, C. Earle Pope, Joelle Abi Rached, Dale Rose, Tiago Saraiva, Shahanah Schmid, Kumiko Shimizu, Sara Shostak, Ingrid Steinberg, Michelle Stewart, Elani Streja, Katherine Thomson, Kalindi Vora, Rachel Washburn, Norton Wise, and Steve Woolgar.

I cannot begin to thank everyone who has helped to inspire and clarify my thoughts on cloning endangered animals through conversation. However, some discussions have been absolutely crucial to my thinking and the subsequent development of this book. Special thanks go to Claire Alexander, Gail Davies, Marie Fox, Tine Gammeltoft, Sahra Gibbon, Paul Gilroy, Donna Haraway, Jennifer Harrington, Claes-Fredrick Helgesson, Stefan Helmreich, Cathy Herbrand, Amy Hinterberger, Klaus Hoeyer, Oliver Hochadel, Martin Johnson, Sharon Kaufman, Susan Kelly, Hannah Landecker, Joanna Latimer, Sandra Soo-Jin Lee, Jamie Lorimer, Rosana Machin, Claire Maris, Jacob Metcalf, Michael Montoya, Lynn Morgan, Manasi Nandi, Sandra Vera Nicolodi, Adriana Petryna, James Porter, Barbara Prainsack, Gisler Priska, Jenny Reardon, Nikolas Rose, Nigel Rothfels, Astrid Schrader, Bob Simpson, Heather Swanson, Mette Svendsen, Karen Sue Taussig, Judy Wacjman, Ayo Wahlberg, Harlan Eugene Weaver, and Zhu Jianfeng.

This project would not have been possible without financial and institutional support from a number of sources. The Department of Social and Behavioral Sciences at UC San Francisco, the Center for Society and Genetics at UC Los Angeles, the BIOS Centre at the LSE, and the Sociology Department at the LSE provided intellectual homes at different times while I was researching and writing this book. In addition, the UC Humanities Research Center supported this research during a crucial time with an Andrew Vincent White Fellowship.

Portions of this book have been published elsewhere. A different version of chapter 1 appeared as "Classification Conundrums: Classifying Chimeras and Enacting Species Preservation" in *Theory and Society* 39(2): 145–172. A portion of chapter 3 appeared as "Transposing Bodies of Knowledge and Technique: Animal Models at Work in Biomedical Knowledge Production" (with Adele Clarke) in *Social Studies of Science* 42(1): 31–52. Isolated sentences and paragraphs are scattered across the book from "Models of Cloning, Models for the Zoo: Rethinking the Sociological Significance of Cloned Animals" in *BioSocieties* 4(4): 367–390.

Last but certainly not least, I am extremely thankful for the love and support of friends and family. I would like to specifically thank my parents Carole and Walter Friese, my siblings Matthew and Sarah Friese, and my extended family, including Alexander, Clarke, and Monica Miller, August, Roman, and Trent Raygor, and Alicia Svenson. This book is dedicated to my partner Stephanie Miller, who has patiently learned about the cloned animals presented in this book with me, always with support, compassion, and love. And of course to our two cats and two dogs, who were there with me during much of the time spent writing this book.

In November 2000 a cloned, endangered bovine was born on an industrial-sized farm in Iowa. Named "Noah," this gaur was created through an interspecies modification in the somatic cell nuclear transfer process, or what is more popularly referred to as cloning. Rather than use scarce gaur eggs in this cloning experiment, researchers from Advanced Cell Technology instead used surplus eggs from domestic cows.[1] Retrieved from a local slaughterhouse, the common genes found in the nucleus of the cow eggs were removed so that the rare DNA in gaur bodily cells could be transferred in.[11] Ultimately forty-four of these novel embryos were shipped to Trans Ova Genetics, a company in Iowa that uses assisted reproductive technologies to selectively breed cattle as part of the beef and dairy industries. Here the embryos were transferred into domestic cows, who acted as the gestational surrogates for their endangered gaur counterparts. In the end, one gaur was born as a result of this experiment in interspecies cloning. Sadly, Noah died just days after birth. But if he had survived, this gaur would have moved from Trans Ova Genetics to the San Diego Zoo in order to become the world's first cloned animal on display in a zoological park.

Amidst high-profile controversies regarding human cloning and human embryonic stem cell research in the wake of Dolly the Sheep, the popular press reported rather positively on the world's first cloned endangered animal. Indeed, one of the surprising facets of this and subsequent endangered animal cloning projects has been the relatively high level of public support for such endeavors. Critics of cloning publicly conceded that even the most nefarious of biotechnologies could find good use. And public opinion polls showed that the American public supported this use of cloning over any other.[3] Reproducing endangered animals is generally met with high levels of support, as conservation has been positioned as an indisputably good thing to do.[4] In this context, the public good associated with reproducing an endangered animal trumped the possibly troubling expansion of cloning across a wider array of species.

Meanwhile, the cloned gaur raised a significant amount of controversy and debate both within the San Diego Zoological Society and across zoos more generally. In direct contradiction to the popularly held notion that cloning endangered animals was an unproblematic and even good thing to do, some zoo people questioned whether this kind of experimentation was appropriate with endangered animals. On the one hand, the incorporation of domestic animal eggs cells and surrogates into the reproductive processes of endangered species raised questions about the classificatory status of cloned animals. Given its biological and social links with domestic animals, could Noah even be considered part of the endangered gaur species? In addition, some questioned how cloning—a technology premised upon creating genetic copies—could ever assist zoos in their attempt to conserve biodiversity. Do cloning and other high-profile, biomedical technologies have a legitimate place in the scientific research agendas of zoological parks? In direct contradiction to public enthusiasm surrounding endangered animal cloning, many people in zoos have been rather ambivalent about such technological developments.

This book describes and characterizes endeavors to clone endangered animals, along with the corresponding controversies in zoos. I contend that different cloned endangered animals embody different imaginaries regarding the future of wildlife on a planet that is increasingly (understood as) shaped by human presence.[5] Nature is thus being innovated in and through cloning experimentation. This is a nature that is marked by an ever-increasing loss of varied habitats, a growing number of species extinctions, and a full range of new kinds of dilemmas posed by global warming. It is also a nature that is being actively made in and through the techniques and knowledge practices of contemporary bioscience and biotechnology. Cloned endangered animals thus provide a window into how nature is being reproduced and transformed within a contemporary milieu that is marked by the joint processes of technological innovation and environmental crisis. As in most areas where reproductive processes are knowingly changed, recalibrating nature with cloning is politically fraught.[6] It is for this reason—rather than a knee-jerk reaction against cloning per se—that the animals produced through interspecies nuclear transfer have been so contentious within zoos.

Reproduction

Somatic cell nuclear transfer has been used to birth animals of five different endangered species worldwide.[7] After the cloned gaur died, Advanced Cell Technology conducted another cloning experiment in conjunction with the San Diego Zoo Global and Trans Ova Genetics. This time, the somatic cells they used were from a different endangered bovine species known as the banteng. The project resulted in the birth of one banteng that is currently on display at the San Diego Zoo.[8] In addition, three litters of African wildcat clones have been born and two unrelated cloned cats reproduced a litter of healthy kittens at the Audubon Center for Research of Endangered Species (ACRES) in New Orleans, Louisiana.[9] At the time of this study, scientists were using the technical process developed with the African wildcats to clone an endangered sand cat. One live birth resulted in 2008, but the cloned sand cat died about sixty days after birth.[10]

While the United States has been the base for much of the activity in cloning endangered animals, similar experiments have also occurred in Europe and Asia. In 2001, scientists announced the birth of a cloned, endangered sheep in Italy.[11] In addition, researchers at the Zoological Society of London initiated a program in cloning amphibians in 2010. Scientists in China have been actively working to clone a giant panda.[12] And in 2008 the birth of a threatened wolf that is native to Korea using interspecies nuclear transfer was announced.[13]

These cloning projects have operated in relation to either the zoological park or the nation-state. The cloned gaur, banteng, African wildcats, and sand cat—along with current attempts to clone frogs—are all embedded in zoos. Meanwhile, the cloned sheep, wolf, and ongoing attempts to clone giant panda are the result of state funding for the scientific production of endangered native species. As such, there are two different kinds of sociological projects occurring under the rubric of "cloning endangered animals," albeit ones that do interconnect.

Social studies of reproduction have emphasized that reproducing people—and, I would add, animals—is interlinked with the reproduction of social orders. The social organization of and cultural meanings attributed to the biological processes involved in reproduction occur in tandem with the reproduction of social forms, such as "the family" and "the nation-state."[14] Reorganizing biological processes through technology thus reproduces and transforms corresponding social orders.[15] State-funded projects in cloning native species are, for example, a site wherein different nation-states are developing their scientific identities as part of a global knowledge economy. These projects open up questions about how national identities are produced and transformed through the technoscientific reproduction of an animal, whose habitat is mapped onto state boundaries.[16] Meanwhile, cloning endangered animals within zoos is a site where ideas about species preservation are being reworked. These projects open up questions about how "nature" is to be reproduced through science and technology. Different endangered animal cloning projects are thus organized with reference to varying social forms, which means that the reproductive work that cloning projects do differs based upon the organization of the experiment. This book thus focuses specifically upon cloning endangered animals in zoos. Research on state-funded projects in cloning endangered native species still needs to be conducted.

Much of the focus in social studies of reproduction has to date been on human reproduction. However, one of the key tensions that reproduction as an analytical device opens up is the relationship between humans and animals.[17] It is well established, for example, that social practices deeming women responsible for reproduction have worked to conceptually equate women with animals and nature. Feminist scholarship has critiqued this equation, showing how an analogy between women and "lower" animals has reproduced gendered hierarchies over time.[18] This conceptual tension is interlinked with a technical tension, wherein the practices of human and animal reproduction are interrelated. It is, for example, also well established that many technologies used to assist human reproduction were initially developed to reproduce animals in agriculture. But where the analogy between different kinds of humans and animals tends to reinforce the separation between humans and animals, the technical link between human and animal

reproduction tends to problematize this divide. The diffusion of reproductive techniques across species has worked to emphasize that the
biological reproduction of humans is continuous with the reproductive
processes of nonhuman animals. This creates a space for exploring the
ways in which the social organization of human and nonhuman animal
reproduction is coconstituted as well.[19]

This book builds upon an emerging area of scholarship that explores
the traffic between human and animal reproduction.[20] This research has
shown that animal reproduction is interlinked with social forms that
are central to human social life, particularly capital, class, and the state.
Extending this research, I ask how the practices of cloning endangered
animals reproduce and transform ideas about nature in a critical historical moment. To carry species forward into the future on a planet
marked by human presence, the biological reproduction of many animals is being transformed. Enacting such biological transformations is
a social practice, a site of human-animal relating. Ideas about nature
and wildlife as separate from humans are necessarily problematized in
bringing endangered animals into being through such relations. I map
how biological transformations in endangered animal reproduction are
coupled with social and cultural transformations in the idea of nature
itself, to which the resulting cloned endangered animal stands as "witness" (Martin, forthcoming).

Biotechnology

It is a truism that much ink has been spilt on the topic of cloning. With
the birth of Dolly the Sheep and developments in regenerative medicine,
cloning reemerged as an ethical problem, one that has been described
as nothing less than "global hyperventilation" (Hartouni 1993).[21] The
prospect of somatic cell nuclear transfer moving from animal to human
bodies has preoccupied much of this discussion. Cloning humans has
generally been assumed to be problematic on the basis of the argument
that somatic cell nuclear transfer creates copies. This mode of reproduction is thought to deterministically create social situations wherein individual identity would be lost, corresponding forms of excessive social
control and commodification would arise, and social relations would be
flattened.[22] Bioethical and popular concern with cloning has thus linked

the social organization of biological reproduction with the reproduction of social orders. However, unlike social studies of reproduction, these discussions have tended to assume that cloning is deterministic and unified in its consequences. I critique this presumption across this book. Countering this dominant narrative on cloning, I argue that, like many other biotechnologies, the meanings of cloning are dependent upon its use and can thus vary across different milieus.

This book continues a long tradition of studying reproduction and reproductive technologies at the interface with Science and Technology Studies (STS).[23] Both fields of scholarship emphasize that technologies are neither deterministic nor value neutral, but are instead sites of practice. Technological practices occur within and are constitutive of social, political, and economic processes that shape the form and meaning of the technology itself. As such, technologies come to embody the visions of their developers, reproducing their interests into the future.[24] But the users of technologies can diverge from the "scripts" (Akrich 1992) that designers create.[25] The goal of STS has thus been to understand technologies as gatherings, or as "matters of concern" (Latour 2004b) that bring different humans and—crucially—nonhumans together.[26] I build upon this field to map out who and what has gathered together in cloning endangered animals, and argue that the form these gatherings take embody an argument for doing species preservation in particular ways. As Donna Haraway (2008: 65) has shown, biotechnologies are ways of "making companions."[27] I ask what kinds of companionships are made in the zoo through cloning practices.

Concepts like actor network theory as well as social worl[d]s nas have been crucial in solidifying the idea that technologies simply reflect nature or society but rather *make* nature and so In other words, science and technology are sites where a range ferent humans and nonhumans come together in productive wa are constitutive of social life. Drawing on this analytic approa book "follows" (Latour 1987) cloning as it has settled down wit ing humans and nonhumans in different zoos. This focus coun notion that cloning determines social forms. Specifically, clor mals are not objects here, but rather embodiments of a particular way to make and sustain social relations.[29] They are "figures" in Haraway's (2008: 4) sense of the word, in that they embody "material-semiotic nodes or knots in which diverse bodies and meanings coshape one another." Social studies of reproduction and science and technology studies combine in such an analytic focus, wherein the cloned endangered animals stand as witnesses to the biological and social relationships through which they were brought into existence. Such an analysis allows us to reconsider the meaning of "cloning" itself by highlighting the unique conditions through which different cloned animals come into being.

A key argument of this book is that the micro- and mesolevel practices involved in organizing cloning experiments serve as models for reproducing nature *into the future*. Different endangered animal cloning experiments have been organized in different ways; the resulting animal stands as an argument for what kinds of relationships should form the basis for remaking nature. This future orientation speaks to both the reproductive and technological aspects of endeavors to clone endangered animals in zoos. On the one hand, these animals embody an attempt to carry forward into the future those animals that are having a difficult time living on this planet, which is marked by an overwhelming human presence. But in addition, much of biotechnological innovation today is rooted in a future orientation that is "promissory" (Thompson 2005).[30] Claims regarding the future use of cloning for species preservation create the grounds for its development in the present.[31]

To conceptualize this dual positioning of a future orientation in cloning endangered animals, I build upon recent anthropological scholarship on potentiality.[32] This emerging literature is centrally concerned with the material and discursive consequences of future orientations that operate in much of biomedicine and bioscience today. Mette Svendsen (2011) has, for example, explored how the potency of embryos to become different kinds of things (e.g., individuals, research materials, organs) is being conceptualized and acted upon in the context of in vitro fertilization (IVF) and human embryonic stem cell research. She contends that, in the cultural landscape of IVF as it is practiced in Denmark, the discourse of waste interacts with the material trajectories of embryos. Svendsen thus focuses on the cultural repertoires that inform how different kinds of futures are enabled in the present. In doing so, she highlights the ways in which language about potential and the material development of embryos are coconstituted. Svendsen's object of inquiry is thereby the developmental process of the embryo. I interpret her work as showing how the medico-scientific space of IVF in Denmark shapes how embryos develop as "material-semiotic" (Haraway 1991) entities in a field of potentiality.

My approach to the future orientations of cloning builds upon Svendsen's articulation of potentiality as an analytic. I also ask how the potential of cloning is being articulated today in practices that are informed by certain cultural tropes. These discourses shape the material development of cells and resulting animals. But where Svendesen focuses on the development of embryos as part of IVF and stem cell research, my focus is the development of endangered animals as part of zoos and species preservation. Where Svendsen's analysis revolves around the question of potentiality per se, I am interested in using this analytic to question how nature is being potentialized in and through the contemporary use of biotechnologies.

Nature

This book builds upon a substantial literature that has explored what biotechnologies do to our notions of nature. These literatures span social studies of reproduction, STS, and environmental studies, as well as environmentalism. I contend that the zoo represents an ideal site for asking how "nature" is being reproduced and transformed through technological means in the current sociohistorical moment that is marked by crisis, in a manner that links these various fields. While the zoo may admittedly be a marginal space in the global development of and traffic in biotechnologies today, it is nonetheless a space that has a long history of mediating changing notions of nature. This aspect of zoos makes it a rich site for considering the meanings and consequences of biotechnologies.

The first known zoo developed in Saqqara, Egypt, around 2500 B.C.[33] Since then humans have variously displayed wild animals in parades, fights, menageries, traveling shows, as well as modern zoos. To conceptualize this heterogeneous history in wild animal collections, Nigel Rothfels (2002) has argued that zoos should be understood as an institution that mediates varied historical and cultural meanings ascribed to wild animals and nature through the social relationships and practices that enable their collection and display over time.[34] While research has explored how the zoo has mediated changing notions of nature through its contemporary display practices, scant scholarship has to date considered corresponding shifts in collection practices.[35] Across this book I show that cloning and other assisted reproductive technologies represent new ways of "collecting" animals for the zoo. Building on Rothfels's definition, I ask how these new kinds of reproductive practices mediate varied historical and cultural meanings ascribed to wild animals and nature. The zoo can thus differently refract the question of nature and biotechnology that has long been of central interest to both social studies of reproduction as well as STS.

Much of environmentalism has been skeptical of technologies generally, including biotechnologies.[36] In articulating the ways in which global warming represents the "end of nature," Bill McKibben ([1989] 2003) vociferously argued against the use of biotechnologies to resolve environmental problems.[37] If the unintended consequences of human technologies left us with a fully altered and thereby "artificial" planet, McKibben ([1989] 2003: 177) warned that biotechnologies would leave us with a planet intent upon human domination. He concluded that this would represent nothing less than the second end of nature.[38]

That said, a more pragmatic approach to biotechnologies appears to be developing in some corners of the environmental movement today. Stewart Brand's *Whole Earth Discipline* (2009) creates a space for technologies within the environmental movement. Given that humans have already changed the planet, Brand contends that the environmental movement must become involved in engineering a more livable planet for a full range of differently situated humans by promoting urbanization, nuclear power, and genetically modified crops. He calls on environmentalists to become involved in developing these technologies to help with the general goals of environmentalism, specifically decreasing the amount of CO_2 going into the atmosphere while creating more space for rewilding.

The debates on the place of technology within environmentalism reflect the divide between humans and nature that has long been a structuring and organizing principle within the environmental movement, particularly in the United States. This idea is currently in crisis, however, evidenced by not only changing planetary conditions through global warming but also by changing ideas about what nature is.[39] In addition, the dubious politics of removing human groups who have long lived in habitats with wild animals has been problematized as a conservation practice.[40] In these contexts, an increasing number of environmentalists are arguing that the environmental movement needs to rethink the relationship between nature and culture, humans and wildlife in pursuing its agendas.

Paul Wapner (2010) has, for example, argued that there needs to be a new way of doing environmental politics, one that appreciates the "otherness" of nature without understanding it as either foundational or as existing outside social, historical, and cultural context. According to Wapner, this means bridging something of an ideological divide. On one side, he positions the environmentalist who values nature as other to humans, as a model upon which human life should be lived, and as a source of aesthetic inspiration. On the other side, Wapner positions "humanists" who see human activity as a source of inspiration, wherein nature is raw material for mastery and control. Rather than forgo both discourses, Wapner contends that conservationists will need to navigate a space in between with a certain amount of ambivalence, uncertainty, and love. Not unlike Brand, Wapner wants to forgo a pure, ideological position in favor of making a more livable planet.

Meanwhile, much of the scholarship in science studies has also explored the question of what "nature" is.[41] These discussions have, however, been more firmly rooted in empirical studies of biotechnologies as opposed to the ontology of the planet in the context of global warming. Such studies have emphasized that the idea of nature as coming before culture developed within a social milieu roughly associated with modernity. This idea was linked with the rise of the sciences as arbiters of Truth.[42] But this binary no longer operates—and possibly never operated—in contemporary scientific knowledge practices. This is because biology and nature are increasingly being made through the social and cultural practices of bioscience, biomedicine, and biotechnology. Nature can no longer be assumed to exist as an a priori condition upon which culture and society are created.[43] Unlike many corners of environmentalism, there is little longing for a "pure" nature within social studies of reproduction or STS, and these arguments have been more or less accepted within the field.

If nature can no longer be understood as a ground for culture, do nature and culture collapse into one another and therein dissolve? Or is a new kind of relationship between nature and culture taking form? On the one hand, some science studies scholars have argued that nature and culture have fully collapsed into one another so that each has lost its analytic salience.[44] On this basis, Bruno Latour (2004a) has argued that ecologists should abandon "nature" altogether. Others contend that the meanings of and relationships between nature and culture are changing. Rather than collapsing into one another and therein dissolving, nature and culture are instead being reworked. Sarah Franklin, Celia Lury, and Jackie Stacey (2000) have, for example, argued that we are seeing processes of denaturalizations and renaturalizations, as "nature" is no longer assumed to be unitary and fixed. This "second nature" is being actively remade. In making this claim, they note Raymond Williams's famous comment that nature does not have a proper meaning (Franklin, Lury, and Stacey 2000: 21) because the idea of nature as essential form and force has long been coupled with the idea of nature as the material world, which *may or may not* include humans (Franklin 2002: 22).

Building upon this later argument, I explore how "nature" is being recalibrated within the zoo through cloning practices. Very few people I spoke with articulated the position that humans are separate from a nature that serves as a unified ground for culture. Some even argued against such a position. Most, however, held onto the idea that "nature" represents the materiality of the planet as well as all the other-than-human beings we live with on this planet. In addition, most people I spoke with also occupied the spaces of zoos and species preservation. And these institutions continue to be highly invested in and organized around older ideas of nature as something that is separate from and prior to human culture, which should guide action. In this context, cloning endangered animals required that people mediate the varied registers that "nature" denotes in a time marked by change and crisis. Drawing on Marilyn Strathern's (1992a) analysis of kinship, the "nature" of these nature preservationists had been made explicitly "mero-graphic," meaning that nature was increasingly understood as part biological and part social.[45] Different cloning projects connected these parts in different ways, and thus engaged in different kinds of world-making projects. Cloning was thus a site where different social practices were used to produce, reproduce, and transform different natures.[46] While it is well-established that we can no longer think of nature and culture as discrete, we know far less about how nature and culture are being actively rethought and reworked in sites like environmentalism and the zoo, which have historically relied upon such a distinction. This book addresses this gap in the literature.

Cloning endangered animals is thus a site where "wildlife" is being reworked as "wild life." Nigel Clark (1999) first delineated "wild life" as part of his attempt to consider the new kinds of imaginaries of nature that have been enabled by both digital and biological technologies.[47] Looking to science fiction, Clark has argued that there is an emerging aesthetic of nature wherein the unpredictability of human-created biological entities is not only understood as potentially destructive but also as generative and creative. Wild life thus articulates the ways in which humans and nature are bound up and coconstituting rather than discrete. But for Clark the word more specifically articulates the ways in which the unruliness of nature is not only a source of fear but also of inspiration, something that humans could and should foster and encourage rather than control and delimit. I build upon Clark's articulation of wild life to consider how cloning is part of the continual remaking of nature, of which humans and technology are increasingly understood as being a part. In the process, I consider how wild life, as articulated by Clark, has itself multiplied and diversified as zoos have continued to mediate new and old meanings ascribed to nature.

Biopolitics

Franklin, Lury, and Stacey (2000: 44) have emphasized that nature has become continuous with "life" in the current sociohistorical moment, a conceptualization that can be dated back to Darwin. In this context, nature has been "reinscribed as the social body in need of management, protection and surveillance" (Franklin, Lury, and Stacey 2000: 44). In other words, as conservation has been based on ensuring the life of the planet, nature has increasingly been understood and acted upon through the logics of biopower, as defined by Michel Foucault (1978, 2003). Across this book I consider the ways in which cloning endangered animals is embroiled in biopolitical regimes. In turn, I use the analytic to reconsider the relationship between humans and animals in cloning projects.

This book thus addresses ongoing debates regarding the applicability of biopower as an analytic within animal studies. Nikolas Rose has alone (2001, 2007) and with Paul Rabinow (2006) sought to clarify and delimit the notion of biopower. Beginning with Michel Foucault's (1978, 2003) well-established articulation of biopower as the productive and generative facilitation of life, Rabinow and Rose have emphasized that this analytic highlights the role of expertise in governing populations, which exerts itself by compelling individuals to conduct themselves in particular ways. This definition emphasizes the connections between Foucault's notion of biopower and governmentality, in that people are governed not (only) through force but also through modes of subjectification. The link between biopower and governmentality has led both Rose (2001, 2007) and Rabinow and Rose (2006) to posit that humans can be the sole subjects of biopower. It is implicitly assumed to be impossible to "conduct the conduct" of animals in the ways described by Foucault; animals cannot be subjectified through expert knowledge regimes.

Feminist science studies have offered a rich set of arguments countering this form of human exceptionalism.[48] Donna Haraway (1989, 1991, 2003b, 2008) has consistently developed theoretical grounds for incorporating nonhuman animals into our understanding of biopower, most recently in her analysis of human-dog relations. Haraway has pointed out, as an example, that the record keeping practices used to create a global industry in animals relied upon the creation of genealogical records, which has been well described by both Margaret Derry (2003) and Harriet Ritvo (1995). Haraway (2008: 53) emphasized that these technologies have also facilitated certain kinds of race- and family-making practices among humans, and have been behind the histories of eugenics and genetics. By illuminating the human-animal-technology interface, Haraway shows that human exceptionalism does not clarify the concept of biopower; it instead erases the traffic in knowledge and techniques used to manage individuals and populations across human and nonhuman animal species.[49] In turn, Haraway rejects the notion that animals lack subjectivity. Rather, humans and animals become together through interactions, which includes those interactions that occur as part of and are authorized by science itself (see also Despret 2004, 2008).

Across this book I consider how cloning endangered animals is embroiled in a biopolitics that links humans and animals through both the traffic in techniques and in corresponding human-animal relations through which both species become together. Endangered animals have been cloned in part because biotechnology companies wanted to find out if they could use interspecies nuclear transfer as part of human embryonic stem cell research. There has been a technical and semiotic link between cloning endangered animals and human embryonic stem cell research. Interspecies nuclear transfer marks in practice the interspecies elaboration of a discourse on "regeneration," wherein bodily and temporal transformations are enacted in the present in order to create new kinds of futures for individuals and populations across the human-animal divide.[50] In the process, humans and animals become together through both the act of cloning animals and through the elaboration of different but interrelated biopolitical regimes. As Sarah Franklin (2007b) has argued in her genealogical analysis of Dolly the Sheep, it does not matter if humans are ever cloned; we are changing ourselves by cloning animals.

Materials and Methods

In conducting the research upon which this book is based, I asked how cloning was varyingly "articulated" (Strauss 1988) within zoological parks in order to clone the gaur, banteng, African wildcats, and sand cat, along with current work in cloning amphibians. In using the word "articulated," I refer to the symbolic interactionist concept "articulation work" developed by Anselm Strauss (1988) and further elaborated with and by Susan Leigh Star (1991). As a sociologist of work and organizations, Strauss developed articulation work to refer to the specifics of putting tasks together in working relation in order to implement and sustain projects. Articulation allowed Strauss to emphasize the interactional processes required to bring different people, who do different kinds of tasks, together. Key to the idea of articulation work is the idea that any project—regardless of its level of standardization—experiences troubles and can break down. How complications are managed, and by whom, is of central concern because these social arrangements indicate how social hierarchies are reproduced over time. Extending this focus, Susan Leigh Star (1991) used articulation work to render the invisible parts of work visible, which is often linked to social inequities.[51] I used articulation work to describe, compare, and characterize the micro- and mesolevel practices involved (and conspicuously not involved) in different endangered animal cloning projects, along with the people, animals, and other things embroiled together (and excluded from) such projects.[52]

To do this, I interviewed twenty-one people who were actively involved in or "implicated by" (Clarke and Montini 1993) endeavors to clone endangered animals in zoos. Five people were interviewed on two separate occasions. This included: 1) reproductive scientists and geneticists working at research centers affiliated with zoos, universities, and biotechnology companies; 2) members of and advisors to Taxonomic Advisory Groups and Species Survival Plans that manage captive populations of endangered animals through the American Zoo and Aquarium Association; and 3) a field conservationist who works on habitat and species preservation in situ. These formal interviews were in some cases supplemented by informal conversations and email exchanges. I also had numerous informal conversations and email exchanges with additional people who were more generally involved in zoos, cloning, and/or conservation, but not formally involved in endeavors to clone endangered animals.

I also visited zoos and affiliated research centers to observe how cloning and related techniques were being used and to see how cloned endangered animals were displayed. I visited the San Diego Zoological Park to see the cloned banteng on display and toured the San Diego Zoo Global's Conservation and Research for Endangered Species (CRES). I visited the Audubon Center for Research of Endangered Species (ACRES) in New Orleans. Here I was able to watch the procedures involved in surgically removing ova from domestic cats and the visual aspects of doing interspecies nuclear transfer. I also visited the related zoo. Finally, I visited the Zoological Society of London, and was provided a tour of the London Zoo. I visited numerous other zoos that have not been actively involved in cloning within the research period.

Finally, I attended professional conferences and analyzed documents pertaining to cloning endangered animals. Here I wanted to see how cloning was situated in larger communities of practice, specifically those focused on reproducing endangered zoo populations as well as developing reproductive technologies. I attended the 2006 annual meetings of the International Embryo Transfer Society, the primary umbrella organization for animal cloning across different species and types of human–animal relations, including endangered species. I also attended the 2006 annual meetings of the Felid Taxonomic Advisory Group that is organized through the American Zoo and Aquarium Association, which is responsible for managing endangered cat populations in U.S. zoos. Throughout the research, I read journal articles, position statements, websites, legislation, book chapters, and newspaper articles pertaining to endangered animal cloning and zoo animal breeding more generally.

I used grounded theory coding (Strauss and Corbin 1998; Charmaz 2000) as well as situational analysis mapping techniques (Clarke 2005) to analyze this material, tracing how cloning projects have been organized and interpreting what these projects mean.[53] In the process, I extended the conceptualization of articulation.

Strauss drew on a definition of "articulation" that refers to a jointed connection, one that permits movement and action. This definition is embedded in anatomy and zoology, and is used to refer to the skeletal structures of animals. It has metaphorically expanded to also describe a conceptual relationship between two different kinds of things (*Oxford English Dictionary* 2009). Physical, sociological, and conceptual joints are often taken for granted, and are thereby frequently made invisible. But when these connections break down, their importance for everyday action often becomes (painfully) clear. This provides a useful metaphor for thinking about the practices involved in project work.

However, I also drew on a second definition of articulation, that of expressing something immaterial and abstract in speech. I coupled a consideration of how cloning is being articulated in practice with the ways people articulate the meanings of those practices in language. A joint never simply exists at the physiological level; it also needs to be articulated in language with a name in order to be known as a joint. Indeed, constructionist approaches to science and medicine have certainly shown that naming things matters. This book is as much about how people involved in cloning endangered animals—including myself—have struggled to articulate the meaning of these cloning practices in words. More than one scientist told me that they enjoyed speaking with me about cloning because it provided a forum to talk about the possible significance of technoscientific developments, for the description of which we do not have an adequate language.[54]

Three Articulations of Cloning: An Overview to the Book

This book explores how cloning has been articulated with endangered animals and species preservation in three different ways to date. By tracing the different practices embodied by different cloned endangered animals, I contend that cloning is a rather "flexible" (Wyatt 2008) technology. As such, the debate over whether or not endangered animals should be cloned in zoos is not about "cloning" per se. Rather, these debates are about what role technoscience should play in reproducing nature at both a physical and conceptual level. The form that cloning projects take is thus deeply contentious, as certain people, animals, and other elements are incorporated, erased, emphasized, and marginalized in the process of making an endangered animal and remaking nature in the zoo.

Chapter 1 sets the stage by introducing the debates surrounding cloning endangered animals in zoos. I do not discuss these debates in general terms. Rather, I locate them specifically in the controversies over interspecies nuclear transfer as a practice, which raises questions about the classificatory status of the resulting cloned animals. This chapter maps out the three major positions taken in my conversations with people regarding this question. One position held that animals produced through interspecies nuclear transfer are endangered species, while another position countered that these animals most definitely are not. There was also a third position, wherein some cloned animals—depending on sex—could strategically count as part of the endangered population. I conclude that these different classifications have been arguments for different scientific practices in zoos. The scientific practices enabled by these different classifications embody different "visions" (Brown, Rappert, and Webster 2000) of the zoo, roles of the life sciences in endangered species preservation, and conceptualizations of nature more generally. Subsequent chapters explore how these different visions are literally embodied by different cloned animals.

Chapters 2 and 3 describe the practices and logics that are being argued for when cloned animals are unequivocally classified as part of the endangered species. The cloned gaur, African wildcats, and sand cat embody this set of classificatory practices. Here cloning is being articulated in order to pursue the scientific identity of the zoological park through technology development, where the spectacle of the cloned zoo animal lies in its technoscientific production. This is based on a set of social practices that I refer to as transposition, which is elaborated upon in chapter 3. Here, the practices and bodies of domestication have been introduced into endangered animal reproduction as a kind of infrastructure. By unpacking this social process, I consider how the cloned gaur, African wildcats, and sand cat are bound up in what Franklin (2006) calls transbiology. Chapter 3 concludes by arguing that this articulation of cloning enables an imagined future in which humans can remake endangered animals so that they can better survive a human-dominated planet.

Chapters 4 and 5 describe the practices and logics that have been argued for when cloned males are classified as part of the endangered species population, but not cloned females. This set of classificatory practices has been premised upon the belief that mitochondrial DNA is maternally inherited. Drawing on this set of beliefs, cloning has been articulated in a manner that directly engages with contemporary practices in reproducing zoo populations. The goal here has not been to physically learn how to engage in cloning, as seen with the gaur and wildcats, but rather to learn how to articulate new techniques with existing practices. In this case, the goal has been to articulate cloning with the kinship charts that are routinely used to reproduce, rather than collect, endangered animals for display in zoos. The spectacle of cloned animals is not directed at the general public here, but rather toward those who manage endangered animal populations in zoos. For these people, cloned endangered animals embody "genetic value" in the kinship charts used to sustain genetic diversity among zoo animal populations. Chapter 5 unpacks the notion of genetic value vis-à-vis the ways in which selective breeding has historically been constituted to create what Harriet Ritvo (1995) has called "genetic capital" in agriculture. Here the goal is to shepherd endangered species and nature in the zoo, which requires privileging genetic definitions.

Chapters 6 and 7 describe the practices and logics that have been asserted when cloned animals are classified as equivalent to hybrids, thereby questioning the role of cloning in zoos. This classification has been embedded in a shift among some reproductive scientists to move their work from technology development toward more basic research on endangered animal physiologies. The goal here has not been to work around endangered animal bodies, but rather to learn how to work with endangered bodies in forging new scientific knowledge and techniques that are species specific. This has not, however, meant that cloning is completely eschewed. Rather, cloning is used as a model system through which reproduction can be understood. The spectacle is not cloning in this instance, but rather the diversity of life forms and biological processes that can be understood through cloning and other techniques. Chapter 7 explores this theme of diversity, focusing on how the environment is increasingly being understood as constitutive of biological diversities as well. The chapter considers how these scientific discourses intersect with conservationist discourse to enable a nature pursued through surprise rather than either control or transformation within the zoo. While this reworks long-standing environmental concerns with the importance of encountering a nature that is other, it also points to ruptures in the genetic values upon which zoos currently rely.

Taken together, this book argues that we need to extend the analysis of how emergent and potentially transformative biomedical and bioscientific practices are being used and envisioned with humans to include those that center upon animal bodies.[55] These animals matter not only because most biomedical knowledge and technologies are first worked out with animals before being transferred to humans. Cloned animals are also worthy of empirical and theoretical consideration because they are embroiled with human counterparts in the dynamic and transformative elaboration of biosocieties, through which nature is being remade from both the inside out and the outside in.

1

Debating Cloning

Cloning endangered animals has been extremely controversial within and across zoological parks. A reproductive scientist who has worked in both the biotechnology industry and in zoos told me about his personal experience with this debate.

> This is an ongoing battle, actually, in the zoo community. I'm still involved in the zoo community, both as a reproductive science advisor but also as a cloner . . . [Cloning] just caused a lot of—well, I go to a lot of the zoo meetings thinking, "Oh, no, what's going happen next?" . . . For instance, one of the people that I followed careerwise, and would love to have worked with, when the [cloned] gaur was done, he stood up at a meeting and said, "No way in hell would this technology ever be useful to the zoo community." And that was kind of like, wow, your idol kind of like crushing your dream.
> *Interview (July 1, 2005)*

Through a somber narration of its personal consequences, this scientist expressed the extent of controversy surrounding cloning in zoos. He wonders and worries about "what's going to happen next" when attending zoo meetings, implying that the debates over cloning are not only ongoing but also continually open to new and potentially contentious iterations. And he has experienced this ongoing debate as challenging at a personal level. His dreams were "crushed" in the process of publicly learning just how negatively his work was received by colleagues and role models in the zoo community.

The controversy surrounding cloning in zoos, made evident in this interview, was a persistent and central feature of my research. Many people I spoke with who were involved in cloning asked that I try to impartially explain why they thought cloning was a reasonable technology to pursue for species preservation. In addition, many people who were opposed to cloning endangered animals asked that I impartially explain why they have decided not to clone endangered animals as part of their research. An occasional person reacted negatively to my even conducting this research, which they thought would give further credence to what they perceived as an illegitimate research trajectory. One scientist told me that my research was symptomatic of all the mass-mediated hype surrounding cloning that he spent much of his time trying to dispel.

Many readers are probably not surprised by this controversy. Indeed, to say that cloning and controversy go hand in hand is an understatement. Dolly the Sheep refueled long-standing anxieties about mass production and the consequent erosion of humanity through technoscience.[1] Shortly thereafter, the debates surrounding cloning became more specialized and professionalized in relation to human embryonic stem cell research in biomedical research and the use of cloning in the agricultural production of food. Debates over the use of nuclear transfer in human embryonic stem cell research centered on the moral status of the embryo, a frame that drew upon the long-standing abortion debates in the United States.[2] Meanwhile, the use of nuclear transfer to clone agricultural animals raised questions regarding the safety of cloned meat for human consumption.[3]

In this context, it would seem to make sense that cloning endangered animals would also be controversial. However, one of the surprising aspects of these endeavors has been the *lack* of controversy in public culture. Indeed, while conducting research I was struck by the generally supportive tone in the U.S. popular press's reporting on the topic. The following statement seemed to sum up the general tenor of the public reaction to cloning endangered animals in zoos:

> Even some critics of cloning say the researchers may have stumbled upon a positive use of the technology. "There are no moral problems with this," said Michael Grodin, a professor in Boston University's School of Public Health who has opposed advances that could lead to the cloning of humans. "There are a host of reasons why cloning humans is wrong, but this could be a positive step toward maintaining these species."
>
> *Associated Press (2000: A1)*

The symbolic value of endangered animals worked to generate support for this use of cloning, with the assumption that reproducing any endangered animal positively contributes to endangered species preservation efforts. Even cloning could find good use. In the context of all this public support, the extent of discord surrounding cloning endangered animals in zoos was rather surprising to me. For me, the controversy was not assumed; it was in need of explanation.

This chapter explores the debates over cloning endangered animals in zoos. Consistent with my argument that the meaning and significance of cloning is grounded in its use, I locate these debates in the ways somatic cell nuclear transfer has been rearticulated with endangered animals to include an interspecies modification. Many people I spoke with were worried by the fact that all endangered animal cloning projects have to date used domestic animal eggs to reprogram the endangered animal bodily cells in the nuclear transfer process. This raised questions about whether or not cloned animals could count as part of the endangered species.

I explore the debates over cloning endangered animals by mapping out how people articulated different positions regarding the classification of animals produced through interspecies nuclear transfer. Reproducing animals is linked with the reproduction of social systems. The kinds of mixtures allowed and disallowed are thereby linked with ideas about how the zoo should reproduce itself as it reproduces endangered animals.[4] As such, the ways people classified cloned animals was related to the kinds of scientific research programs they wanted zoological parks to pursue into the future. These debates were thus not simply contests over resources and prestige for a particular laboratory and the kind of scientific research therein pursued. More broadly, contestations over interspecies nuclear transfer mattered because different scientific knowledge practices constitute "nature" in different ways. Whose knowledge practices will shape the "material-semiotics" (Haraway 1991) of endangered species into the future lies at the heart of the debates over cloning endangered animals.

Articulating Interspecies Nuclear Transfer

Somatic cell nuclear transfer works by fusing a somatic cell into an enucleated egg cell in order to create an embryo that has the same genetic makeup as the somatic cell donor. Here, the nucleus of the cell comes from one individual, while the cytoplasm comes from a different one. This process seeks to recapitulate the genome of the somatic cell donor exactly, hence the notion of the copy. However, there is mitochondrial DNA in the cytoplasm that comes from a different individual.[5] Marilyn Strathern (1992b: 6) has commented that neither biological nor social reproduction ever results in forms or processes that are exactly the same, and this rather ironically also includes cloning. In the context of interspecies nuclear transfer, this difference is important because the mitochondrial DNA, inherited from the egg donor, comes from an animal of another species. In the contemporary parlance of the life sciences, these embryos and resulting animals are referred to as "heteroplasmic" because mitochondrial DNA is inherited from a different, domesticated species. These animals are also characterized as "chimeras" in everyday language, according to traditional definitions of the word.[6] In the United Kingdom, these entities are also referred to as "chybrids" in the popular press and policy papers. For this chapter, I will use the word chimera as it is an idiom that cuts across different social groups and is thereby most likely to be recognized by the greatest number of readers.

Chimeras produced through interspecies nuclear transfer are difficult to categorize within the existing classification systems used to distinguish species.[7] The biological notion of species assumes, as its primary mechanism, biparental sexual reproduction (Claridge, Dawah, and Wilson 1997). The hybrid has long been viewed as problematic to this notion of species because these animals represent sexual reproduction between two different kinds of animals, resulting in a genetic mixture of DNA in chromosomes of the nucleus. Chimeras are similarly troubling as genetic mixtures. However, the mode by which these mixtures are produced creates an additional challenge when classifying these biological organisms. Whereas hybrids are sexually produced mixtures in nuclear DNA, chimeras are asexually produced genetic mixtures at the cellular level. As such, the mode by which genetic mixtures are produced in chimeras is not accounted for by the species notion that is biased toward sexual reproduction.[8] Here, the nuclear DNA is from one species while the mitochondrial DNA is of another species as a result of asexual reproduction.

Chimeras pose a kind of classification conundrum to the biological definition of species, which in turn poses a set of problems for zoos that are engaged in species preservation. Species distinctness is crucial for the official classification of "endangered species," which relies on the scientific division of different kinds of animals. In this context, genetics works to naturalize the categories of both species and endangered species, enabling the development of social orders that support the preservation of these animals.[9] Counting as part of an "endangered species" matters because this official classification provides certain species with significant legal protections from physical harm. Genetic contributions from both an endangered animal and a domestic animal call into question the legitimacy of these protections.

In this context, many people I spoke with sought to determine what chimeras are and how they should be classified by considering the biological significance of mitochondrial DNA. However, this common reference point did not consolidate the debate.[10] Rather, people made mitochondrial DNA meaningful in different ways by emphasizing particular biological facts and marginalizing others in reference to culturally available categories.[11] In this context, people engaged in the debates over cloning endangered animals by articulating particular positions on what cloned animals are and how they should be classified.

To examine these debates, I used positional maps to analyze the contestations surrounding the classification of chimeras. This is an analytic strategy that Adele Clarke (2005) developed as part of her situational analysis. Here all positions regarding an area of interest, concern, or controversy are mapped vis-à-vis two key discursive elements in the positions themselves. Figure 1.1 is the map I made of all the positions people took regarding the classification of chimeras, which are organized according to two axes: a) the significance of mitochondrial DNA to species boundaries, and b) the degree to which animals produced by interspecies nuclear transfer count as part of the endangered species in question. I chose these axes because they reflect the key question (e.g., are these cloned animals endangered species?) and the primary referent (e.g., mitochondrial DNA) in the debate. This map allowed me to lay out all the different positions I heard articulated regarding what cloned animals are and how they should be classified.

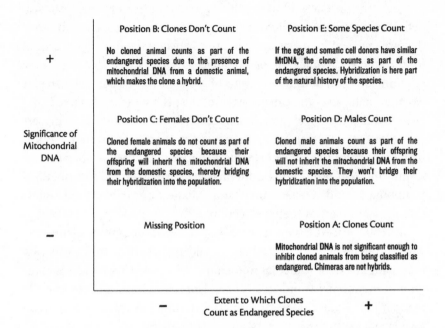

Figure 1.1. Positions on the classification of cloned animals.

With positional maps, Clarke (2005) provided a methodological intervention through which debates can be understood as discourse elaboration rather than (or in addition to) group-based politics, wherein different groups' interests are at stake. In this context, and to further consider these positions as discourse, I asked what and how different schemas were used to articulate these varying positions in language. Schemata are the familiar images that people use to cognitively process new information.[12] Sociologists and anthropologists have emphasized that schemata do not result from innate cognitive processes, but are instead collective and socially mediated.[13] "Hybrids" and "bridges" provided important reference points as people sought to understand and articulate what cloned animals, produced through interspecies nuclear transfer, are in relation to endangered species.

Through schematic references, things get lumped together by ignoring differences or split apart by exaggerating differences in ways that are socially informed rather than ontologically determined.[14]

Using a particular schematic reference, different people emphasized different biological aspects of chimeras while marginalizing other biological facts. As such, people had to weave culture and nature, biology and society together in order to render cloned animals meaningful. How people did this linguistic articulation work was related to the kinds of preservation practices that they wanted to see implemented in the future.

Figure 1.2 is my revised positional map, representing a kind of "zoo map" that structures the remainder of this chapter and book. Where the zoo map allows visitors to find the animals they want to see, this map allows me and readers to find their way through the material and semiotic situations of cloning endangered animals in the zoological park. This map is structured by showing which schemata informed the different positions described in Figure 1.1. The map also shows what the consequences of each linguistic articulation of cloned animals are for preservation practices.

Talking about Chimeras

The classification conundrum posed by animals created through interspecies nuclear transfer was rendered familiar by referencing existing categories, or familiar schemata. Hybrids and bridges were commonly used as metaphors for understanding what clones are and their consequent species relation. I will briefly review what hybrids and bridges mean in the specific subcultures of zoos and species preservation before showing how these categories were used to classify cloned endangered animals.

Chimeras as Hybrids

Almost everyone I spoke with used hybrids as a reference point in rendering chimeras meaningful. Hybrids are animals produced through sexual reproduction between two different species in the same genera, wherein the genetic mixture lies in the nuclear DNA of the individual. As already discussed, clones are heteroplasmic biological organisms produced through asexual reproductive processes. Here, the nuclear DNA is derived from one individual alone. However, mitochondrial

Figure 1.2. Zoo Map: Classifying Cloned Animals and Enacting Species Preservation

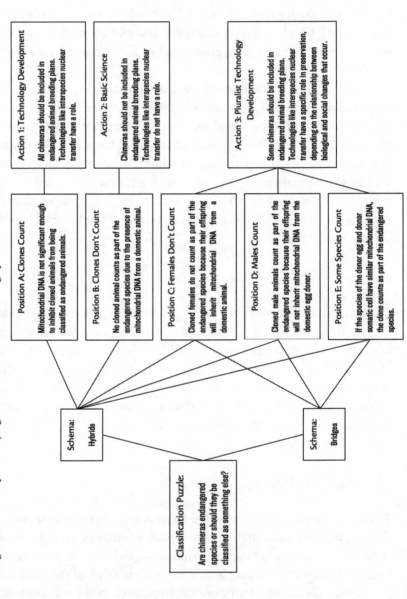

DNA in the cytoplasm comes from a different animal. As such, hybrids and clones are both biological organisms that are distinguished as interspecies genetic mixtures. However, the kind of genetic mixtures these organisms represent are substantially different. Nonetheless, it is important to understand what hybrids mean within zoos and species preservation more generally.

Hybridization has long been considered an unnatural, evolutionarily inferior mode of sexual reproduction because it is believed that the resulting offspring are less fit to reproduce. The mule, which results from hybridization between a horse and a donkey, is a classic example because these animals are sterile. This point is clearly made in an article by Donald Levine (2002: 254), which addresses the dangers of hybridization for endangered wildlife:

> [D]ifferent species in the same genera do not normally interbreed . . . Although such hybridization never takes place in the vast majority of genera, it is quite common in some. . . . With hybridization so rampant, one wonders how species ever maintain their distinctness. They do, in part, because the production of hybrids does not necessarily shift genetic material between species. For genes to traffic in this way, hybrids must cross with at least one of the parent species. In many instances that doesn't happen. Why? As Darwin had observed, most hybrids are inferior to their parents.

The inferiority of hybridization in evolutionary terms works to naturalize the notion of species as distinct, sexually reproducing populations.

However, there has been increasing recognition that hybridization is not always an inferior mode of sexual reproduction. At times hybrids will outperform their parent populations in ways that can dramatically change landscapes and ecosystems.[15] Hybrids become dangerous here because they can lead to the extinction of the parent species. Indeed, while touring zoological parks, I commonly found that hybridization is a source of endangerment for an array of animal species, including the banteng. These hybrid populations are still understood as unnatural, however, because their genetic admixture is often related to human intervention, specifically transportation systems that move plants and animals around to different parts of the world.

Given that hybridization is considered an inferior and dangerous mode of reproduction, it is not surprising that hybrid animals are often considered less valuable than animals with a lineage in one species or subspecies among zoological parks.[16] A reproductive scientist demonstrated this point by telling me about the decision to "phase out" hybrid lions from U.S. zoos (Interview, April 11, 2006). All institutions were asked to stop breeding these lions and today the U.S. captive population is solely made up of African lions. This researcher commented that the current lion population is considered more valuable because it is pure.

But hybridization and hybrid animals are not always considered problematic in conservation and zoo worlds. The same researcher followed the narrative regarding hybrid lions with a discussion of the preservation plans for the Florida panther. Here is a case where hybrid animals and hybridization have been rendered valuable. The U.S. Fish and Wildlife Service (1993) has an ongoing plan to recover in situ, self-sustaining populations of Florida panthers by protecting habitats, reintroducing panthers to these habitats, and engaging publics so that the panthers are not harmed. However, the Florida panther population was so small that, to make it sustainable in situ, new genetic information had to be incorporated into the population. Specifically, more genetic diversity had to be created by hybridizing the Florida panther with the Texas puma. What made this kind of hybridization acceptable, while other forms of hybridization are not, was scientific evidence that the Florida panther had at one point in time dwelled across North, Central, and South America in coexistence with the puma. Thus, the two species have a history of cohabitation and hybridization. The difference between Florida panthers and Texas pumas is a biological artifact of social processes based on human intervention. What this case shows is that hybrids can be used to reanimate what was once in existence before human interference. But hybridization cannot to be used to create something new through human intervention. Hybrids embody the long-standing belief that human activity is cultural and can be separated from the natural world.[17]

The image of hybrids is thereby very much intertwined with domestic animals and domestication. Zoos, as sites that display "wild" animals, are not meant to display domestic animals, or animals created through human invention. Hybrids are more or less treated as equivalent to domestic animals in this institution. These are animals that result from human intervention and are thereby different from the wild animals that zoos and conservationists are trying to preserve.

Chimeras as Bridges

The image of the hybrid was at times supplemented and modified with the image of a bridge. The bridge conjures up the idea of an infrastructure that allows one to bypass treacherous terrain to arrive at a desired destination more easily. In order to explicate how this image was used to classify heteroplasmic clones, we need to turn to mainstream practices in ex situ species preservation.

The American Zoo and Aquarium Association developed Species Survival Plans (SSPs) during the 1980s to selectively breed individuals in the captive population in order to maximize genetic diversity. Kinship charts and studbooks are the primary technologies used to determine which animals should breed and which should not in order to maximize the genetic diversity of the captive population. These organizations will be discussed in more depth in chapters 4 and 5, but for now it is important to note that hybrid animals are not included in these record-keeping and population management practices. Hybrid animals are thereby erased from the technical and physico-genetic body of the captive animal populations in U.S. zoos. If cloned animals are classified as hybrids, these animals will also be excluded. Interspecies nuclear transfer will literally not matter in the efforts to preserve endangered animals.

Using the image of a bridge, however, it becomes possible to integrate interspecies nuclear transfer into Species Survival Plans while addressing concerns raised through schematic reference to hybrids. One of the primary goals in cloning endangered animals has been to "rescue" (Ryder and Benirschke 1997) lost genetic information in order to create genetic diversity in these small populations. Cloning represents a way to do this kind of work, making the cloned animal a kind of "bridge" individual. Because it is believed that mitochondrial DNA is passed through the generations by females alone, males could provide a way to create a bridge that brings in valued genetic information from an endangered animal while keeping at bay devalued genetic information from a domestic animal.

The image of a bridge has precedence in zoos with the practice of interspecies gestation. Embryo transfer made it possible to "contract out" the reproductive labor involved in gestation to animals of less genetic value, including closely related domestic species. Shellee Colen (1995) coined the term "stratified reproduction" to refer to the hierarchically differentiated values attributed to reproduction, wherein biological reproduction is more valued than social reproduction. Extending this concept, the biological processes of reproduction are also stratified. The reproductive labor of genetic inheritance is deeply valued whereas gestation is devalued. This coincides with moves in biomedicine more generally that classify gestation as part of "nurture" and therein "society," whereas fertilization is seen as a biological process that is part of "nature."[18] Positioning gestation as a social relation deems interspecies gestation unproblematic when used with endangered animal bodies. Here, the categorization of organisms reproduced through interspecies gestation has been firmly positioned within the endangered species from whom the embryo was derived. The reproductive labor of the domestic animal has thereby been effectively erased in the classification of the resulting individual. Domestic animals thereby have a history as bridges that allow for the reproduction of endangered animals.

Classifying Cloned Animals

So far I have discussed why interspecies nuclear transfer creates animals that are heteroplasmic or chimeras, containing nuclear DNA from an endangered animal but mitochondrial DNA from a different, domesticated species. I have also situated the significance of these chimeras in both biological and sociocultural terms. Having provided this background, I now explore the debates over cloning in zoos by looking at the different positions people have taken on the classification of cloned endangered animals.

It is important to emphasize that I do not think there is a right or wrong answer to the question of what cloned endangered animals are and how these animals should be categorized. In their study of classificatory practices, Geoffrey Bowker and Susan Leigh Star (1999: 7) have pointed out that any "category valorizes some point of view and silences another. This is not inherently a bad thing—indeed it is inescapable. But it is an ethical choice, and as such it is dangerous—not bad, but dangerous." With this in mind, I am not trying to valorize or demonize one set of classificatory practices over another. Rather, I want to ask what is remembered and forgotten (Bowker 2005; Bowker and Star 1999) or highlighted and hidden (Lakoff and Johnson 1980) in the different positions that varyingly lump and split (Zerubavel 1996) clones in relation to different species. In doing so, I show that highlighting and hiding certain aspects of cloned animals in relation to endangered species works to prioritize a particular way of doing science and conservation in zoos while marginalizing another. The different ways of doing science in zoos, and the consequences those practices have for the ways nature is made, are the focus of the remainder of the book.

Position A: Clones Count as Endangered Animals

One prominent position regarding the classification of chimeras holds that these organisms can be classified as part of the endangered population. This position was articulated in everyday and official discourses. The everyday language of "cloning endangered animals" holds that mitochondrial DNA is not significant enough to inhibit cloned animals from being considered part of the endangered species. This language was used across popular cultural texts, scientific reports, and in the shorthand language of people involved in this endeavor. Here the cloned animal was lumped together with the somatic cell donor and corresponding endangered species. The dominant language of "cloning endangered animals" assumes that the somatic cell donor is the subject of the reproductive process. This language ignores the involvement of an egg cell donor and the corresponding inheritance of mitochondrial DNA from an animal of a different, domestic species.

This position is assumed in everyday language and reflects the way genetic reductionism is embedded in the "automatic cognition" (D'Andrade 1995; DiMaggio 1997) of U.S. cultural landscapes.[19] Indeed, genetic reductionism is acutely present in contemporary cloning discourses and practices. Sociologists and anthropologists have argued that the notion of a clone as an exact and complete replica reveals the extent to which nuclear DNA is understood as the basis for personhood and individuality.[20] It is implicitly assumed that the genetic information found in the nucleus of a cell will reanimate an endangered animal in its entirety. Significantly for the case of interspecies nuclear transfer, this version of genetic essentialism relies on the long-standing assumption that nuclear DNA is the foundation of the body in question, while the mitochondrial DNA in the cytoplasm of the cell can be ignored as insignificant.[21]

While the everyday language of cloning endangered animals forgets mitochondrial inheritance unknowingly, some people involved in or implicated by these cloning practices explicitly contend that interspecies mitochondrial DNA inheritance can knowingly be forgotten in official classification systems. The specific official classification systems in question here are the records zoos use to track their captive, endangered populations and manage their reproduction. This position was articulated as follows:

> I personally, I guess, will have to decide—at some point—whether or not I want to keep this banteng in the Studbook and use him when analyzing the population. And he does, phenotypically he is a banteng.
> *Interview, Species Survival Plan manager (October 18, 2005)*

> In terms of having the mitochondria and the mitochondrial DNA from a different species, it's just that. I think it's a trade-off. I think if you could get around it you would, but if you cannot it's a reasonable approach. We certainly understand that many inherited diseases and many fundamental metabolic processes are mitochondrially driven, and so it's true that you won't be able to recapitulate the species exactly as they are in the wild. But, still, I think that's a reasonable approach to solving the problem.
> *Interview, University scientist (April 20, 2006)*

> Are you better to preserve 99.99 percent of that genetic lineage and tweak a little bit and let it continue? Or are you a purist enough to say

"No, it was going out anyhow. You shouldn't genetically try to tweak it. Just let it go." I think most people would probably say, in the final analysis, preserve 99.99 percent of the lineage; do a little bit of tweaking.

Interview, Field conservationist (April 25, 2006)

At present, this position is officially held in the display practices of the San Diego Zoo, where a cloned animal is labeled as a banteng, and in the North American banteng studbook.

The significance of interspecies genetic contributions is diminished in these statements by emphasizing the similarities between the cloned banteng and the banteng species. The chimera looks like a banteng and so, drawing on the long-standing use of morphology in classifying species, can be lumped together with the endangered species. While chimeras may have differences in inherited disease and metabolic process, these differences are not the basis for subspecies differentiation. The genetic changes produced through chimeras are deemed insignificant by many in the context of species extinction. Being a purist comes at the expense of extinction. In this position, clones are like the endangered species and different from hybrids.

Position B: Clones Do Not Count as Endangered Animals

Other people articulated an opposing position that countered the everyday language of cloning endangered animals as well as those who argue that mitochondrial DNA is insignificant for the official classification of chimeras. This position holds that no animal produced through interspecies nuclear transfer should be considered part of the endangered population and was expressed by reproductive scientists working at zoos and Species Survival Plan managers.

In my opinion, as the coordinator of an endangered cat SSP, I would not let that [cloned] animal be part of the managed population because it's not an endangered cat any more. It's something else. And it's just like if I got a fishing cat–domestic cat hybrid. I'm not going to manage that in the population.

Interview, Zoo scientist and Species Survival Plan manager (April 8, 2006)

If you track the [cloned] animal, now the DNA of the mitochondria is from a domestic cow. So we feel that by doing that you're basically forming a chimera. You no longer have a pure strained banteng because the mitochondria are from a domestic cow whereas the nuclear DNA is from a banteng. So right there we run into problems and these may spill over, potentially, into legal and legislative issues. Is it a banteng? If people can argue it's a chimera, it no longer becomes a protected species. So there are political aspects for us to at least be aware of.

   Interview, Zoo scientist (April 11, 2006)

This position has garnered official status through the U.S. Fish and Wildlife Service, which has classified the cloned banteng as a hybrid.

We again see genetic reductionism shaping the image of the chimera, but with a very different position resulting. Here genetic purity in both the nucleus and the cytoplasm of the individual is of immediate concern and interspecies mitochondrial inheritance cannot be forgotten. Whereas lumping entities together requires ignoring certain differences, differentiating entities often requires exaggerating variations. The differences between chimeras and the endangered population are exaggerated in this position, thereby effacing the similarities between chimeras and the endangered species that were discussed in the previous position.

This position was also articulated vis-à-vis standard practices used by geneticists for determining the origins of species, often traced through mitochondrial DNA. Whereas nuclear DNA represents the combination of maternal and paternal inheritance, mitochondrial DNA is believed to be inherited through the egg and hence is maternally determined. Mitochondrial DNA is thus used to determine genetic founders by tracing a population back to one female. These tests are also used to determine the species and kinship relations in wild populations of endangered animals.[22] On the basis of this technical practice, some have raised concerns about the consequences of interspecies nuclear transfer for these categorizing practices. Somewhat frustrated that this concern was not being more thoroughly acknowledged in discussions about cloning, one geneticist working in a zoo commented to me: "If anybody is going to trace the lineage of a certain individual, if they have to go back looking at mitochondrial DNA, I think they've got a mess on their hands with this situation" (Interview, April 11, 2006). According to this technical practice in genealogy tracking, a clone would be reclassified as a domestic animal because its mitochondrial DNA was inherited by the egg cell donor. Meanwhile the inheritance of nuclear DNA from a different, endangered species would be hidden.

Positions C and D: Male Clones Count as Endangered Animals, but Females Do Not

A middle ground approach to the polarized positions presented above is seen in the argument that male clones can be classified as part of the endangered species but females cannot. As previously discussed, mitochondrial DNA is believed to pass through the generations by way of the egg cell alone.[23] Drawing on this assumption, it was posited that interspecies mitochondrial DNA is quarantined in the male chimera's body and so males can count as part of the endangered population. Meanwhile, female clones cannot be classified as endangered because their compromised genome risks leaking into the population. These positions are exemplified in the following articulations:

> Well, according to the U.S. Fish and Wildlife the mitochondrial DNA from the domestic cow makes the cloned cow a hybrid, which is good for us because the animal doesn't count as a banteng and so doesn't require permits. But males don't pass mitochondrial DNA to offspring so hybridization should be gone with the next generation. And this circumvents any criticism that cloning dilutes the gene pool.
> *Interview, Zoo scientist (July 1, 2005)*

> You might say cloning could be beneficial for males, but is less pure and less efficient for females. The charge that I've heard leveled about this is that, for example, the banteng or the gaur was actually a hybrid. That's not technically true. It's not a hybrid, but certainly is contaminated with mitochondrial DNA. To what extent the male banteng is contaminated with mitochondrial DNA is unknown, but is certainly significantly less than if the offspring was female.
> *Interview, Zoo scientist (May 16, 2006)*

I do not think we should be cloning females. Now, the males can act as bridges to bring in new [genetic] information to the captive population. And, we don't know this for sure, but we believe they won't pass on their mitochondrial DNA to their offspring. But females will and so, I don't think we should be doing that. Those females can't be part of the endangered population because they are going to spread the mitochondrial DNA from domestic animals.

> Field notes, informal statement made by an industry scientist at the International Embryo Transfer Society Meetings (January 8, 2006)

This position has garnered official status in the Studbook and Species Survival Plan for the banteng, which includes the cloned animal in its records and breeding protocols. To date, no female chimera has been excluded, largely because there has been no case to debate this position. This is because the African wildcat is not a managed species in U.S. zoos.[24]

These positions are informed by a genetic definition of species that incorporates both the nucleus and the cytoplasm, like the position that no cloned animal counts as endangered. But the focus here is not on the hybrid as an individual. Rather, hybridization as a biological process is highlighted. It is argued that chimeras may be like hybrids in that they are genetically contaminated, but it is their capacity to hybridize that shapes their classificatory status. The biological process of forgetting interspecies mitochondrial DNA justifies the social forgetting of this information in the classification of male chimeras as part of the endangered population.

Position E: Whether Clones Count as Endangered Animals Is a Species-Specific Question

Finally, one zoo scientist I spoke with articulated the position that the significance of mitochondrial DNA is species specific. Here, the geneticist was drawing upon emerging findings in the field of population genetics, which show that some species have distinct nuclear DNA but similar mitochondrial DNA. This was referred to as cytonuclear discordance in the interview. Because the cloning process and creation of chimeras matches the natural history of some endangered species, this position holds that some heteroplasmic individuals can be categorized as part of the endangered population. Not unlike the position that male but not female clones can be categorized as endangered, this position argues that alien mitochondrial DNA can be forgotten in some cases. Here, the socially circumscribed genetic relations produced through interspecies nuclear transfer match the genetic marking made by natural history. As such, alien mitochondrial DNA can be forgotten in the classification of some clones. The natural history of a bridge between two subspecies justifies the social and technical enactment of the same bridge for preservation purposes, which is similar to the logic of hybridizing the Florida panther with the Texas puma according to U.S. Fish and Wildlife. To my knowledge, this position has not been enacted in the official classification of clones.

Missing Position: Chimeras Are Not Endangered Animals, but Not Because of Their DNA

It is important to point out that there is a possible position regarding the classification of chimeras that was not articulated. This position would go something like this: cloned endangered animals cannot be considered part of the endangered species population, but for reasons other than their mitochondrial DNA. On the one hand, this missing position alerts us to the salience of genetic markers in delineating endangered species. On the other hand, the missing position forces us to ask who might articulate this position. Who might we need to speak with in order to hear this position articulated?[25]

While this position was not explicitly articulated in relation to questions regarding the classification of chimeras, some study participants made comments during interviews that provided insights into how this position might be articulated. Here are two examples:

So how do they draw a distinction, then, between cloned animals of endangered species and domesticated dogs and cats, which are wards of human goodwill and interest? We've genetically modified them down the ages, albeit not by high tech means. But essentially you end up with the same thing. You end up with animals that are domesticated, and are there by virtue of our having taken over their genetic futures.

Interview, Field conservationist (April 25, 2006)

For a while it was so popular and so sexy to do all these assisted reproductive technologies. We were thinking, "Well, gosh, we'll never have
to have animals together any more. We can do it all artificially." Well, if
our eventual goal truly is reintroduction or self-sustaining populations
that doesn't—those two things don't work. You don't want to produce
a whole generation of animals that can't reproduce on their own. It
makes no sense.

Interview, Zoo scientist (May 16, 2006)

Drawing on these statements, I propose that the missing position
would articulate concern with chimeras not as a genetic entity, but
rather as a relational entity. While some exclude clones from the endangered population on the basis that their genome is domestically compromised, the missing position would exclude chimeras on the basis
that these animals come into being through a relation with humans that
is rooted in control. And here we begin to see that "endangered species"
is not simply a biological category; it is also a very particular kind of
relation that humans forge with animals.

Enacting Endangered Species Preservation

Debates over the classification of clones exemplify the processes
involved in delineating new biological entities in the context of existing social orders. Chimeras create a problem for the logic of species,
which has been dependent upon biparental sexual reproduction as the
primary mechanism for speciation. On the one hand, these organisms
can be ignored as aberrations in favor of sustaining existing classification systems. On the other, these organisms can be embraced in favor
of creating new conditions of possibility. In her classic text on classification, Mary Douglas ([1966] 2005: 117) provocatively points out that
those entities difficult to classify have potentiality precisely because of
their destructiveness:

Granted that disorder spoils pattern, it also provides the material of pattern. Order implies restriction; from all possible materials, a limited
selection has been made and from all possible relations a limited set has
been used. So disorder by implication is unlimited, no pattern has been

realised in it, but its potential for patterning is indefinite. This is why, though we seek to create order, we do not simply condemn disorder. We recognise that it is destructive to existing patterns; also that it has potentiality.

All the positions regarding the classification of chimeras are linked to certain actions that are directly relevant to the social and material order of endangered species and their preservation. Clones are disorderly, but they provide material for new relations and patterns that may assist and/or damage endangered wildlife. Cloning becomes a means for mediating older meanings of nature as fixed with new technological possibilities, so that nature can be reproduced—albeit as a different kind of thing—into the future.

I turn now to the actions that result from varying positions and corresponding schematic references. The classification of clones informs the practices of science and preservation in a manner that has direct consequences for the meaning and materiality of endangered animal bodies.

Action 1: Allow All Clones in the Breeding Population

If one understood clones as different from hybrids, the person would often go on to argue that these animals should be classified as endangered. This understanding was premised upon the assumption that the DNA that matters for speciation is in the two sets of chromosomes within the nucleus of the cell. Here, mitochondrial DNA either did not matter or played a role that is not relevant to the question of species boundaries. The position that all cloned animals count as endangered animals held the schema of hybrids and the act of inclusion together. The goal here was to use new technologies and readily available domestic animal bodies in order to make endangered animals. This position thus seizes the opportunities presented by chimeras for changing the biological order of endangered species, in order to make these species more likely to survive on a changing planet. This action is embodied by cloned endangered animals produced through the logic and practices of technology development, a mode of conducting science in the park that will be discussed in chapters 2 and 3.

Action 2: Allow Some Clones into the Breeding Population

If one understood clones through schematic reference to hybrids and bridges, the person would normally go on to argue that some can be classified as endangered while others cannot. This was either a question of sex or a question of species. Such an understanding of cloned animals was premised upon ex situ preservation practices based on managing the genetic configuration of captive endangered animal populations using selective breeding techniques. Here the schemas of hybrids and bridges are held together with the act of partial inclusion. The goal here has been to create genetically diverse, captive populations of endangered animals in the context of changing collection practices among zoos.

It is important to put this position and corresponding set of practices in context. Zoos historically relied upon the wild animal trade to display animals, which resulted in high death tolls and contributed to the endangerment and extinction of some species. In order to rectify this history, most zoos today want to create "self-sustaining" (Benirschke 1986), captive populations through managed breeding, wherein genetic diversity is the goal. However, these efforts have been hampered by the need to bring new genetic information into the captive population so that it doesn't become inbred.

Cloning represents an opportunity to address this problem by articulating sexual and asexual reproduction. In her genealogy of Dolly the Sheep, Sarah Franklin (2007b) notes that Ian Wilmut developed somatic cell nuclear transfer with the explicit assumption that combining sexual and asexual reproduction carries the greatest evolutionary advantage. This assumption has traveled to zoos, at least implicitly. Combining asexual reproduction with existing selective breeding practices allows zoos to bring in new genetic information without having to remove wild animals from their habitats. Stores of cryopreserved cell lines from deceased animals can be asexually reproduced. Extending this, tissue samples can be collected from wild animals as opposed to removing entire animals from their habitats. Where those seeking to keep all clones within the endangered population are centrally focused on transforming the biological order of endangered species, this position is centrally focused on changing the social order of zoos.

Here zoos are trying to change their collection practices while also delimiting the biological changes that are thereby made to endangered animals as a result. Everyone involved in the study wanted to change the materiality of endangered species by making these populations bigger, more genetically diverse, and thereby sustainable into the future. But those who understood chimeras through schematic reference to hybrids articulated concerns about transforming the genetic composition of these populations in the process. The schematic combination of hybrids and bridges addressed this concern, while also articulating a space for interspecies nuclear transfer when used with males or species bodies that have a natural history of bridging.[26] Such practices are embodied by the cloned banteng, which will be explored in chapters 4 and 5.

Action 3: Do Not Allow Clones into the Breeding Population

If one understood clones as the same as hybrids without reference to bridges as an additive schematic reference, often the person went on to argue that these animals should not be considered endangered. This understanding of clones was based on a tradition of excluding genetic mixtures produced through hybridization from legal protections offered by the category of "endangered species." This position held the schema of hybrids and the act of exclusion together. The goal was to preserve endangered species as a legal entity by sustaining their bodies as they exist today, which is believed to be independent of human affairs. If wild and domestic cows are not sufficiently different, how are life-saving protections for the former sustained while the latter can be eaten for dinner?

Lumping chimeras with hybrids means that cloned animals will be excluded from the breeding population and their genome will end with their life course. Not unlike the hybrid lions, these cloned individuals will literally be forgotten and erased from the zoo population. For some this action and corresponding material consequence was considered positive. One of the charges against the language of "cloning endangered animals" was the erasure of changes made to endangered animal bodies through interspecies nuclear transfer. These cloned animals embody a novel genetic composition, providing a mark of their social creation. This genetic composition not only changes the resulting animal, but potentially the population more broadly through their reproduction. As such, both chimeras and hybrids were deemed problematic because of their "unnatural" creation. In the process, interspecies nuclear transfer was deemed irrelevant to ex situ endangered animal preservation.

Many people who took this position actively decided not to clone endangered animals. However, others were nonetheless using cloning in experiments with domestic animals as "surrogate" (Bolker 2009) models of endangered animal physiologies. This set of classificatory practices will be explored in chapters 6 and 7, where the focus is on understanding the diversity of biological forms and processes amongst an array of endangered species.

Conclusion

Based on my experiences in endangered species preservation worlds, it appears that there was a temporal component to the development of the different positions vis-à-vis one another. The position that no clone can be classified as an endangered animal seems to have arisen in reaction to the everyday language of cloning endangered animals, which was garnering a significant amount of positive media attention. Some individuals indicated that schematic reference to hybrids served a rhetorical function here, in that this image called into question both the classification of chimeras and the usefulness of interspecies nuclear transfer to zoos. Some responded by rearticulating their position that cloned endangered animals do count as endangered animals, and rejected schematic reference to hybrids on a technical basis. However, others sought to account for concerns about genetic impurity while simultaneously arguing for the usefulness of interspecies nuclear transfer to ex situ endangered animal preservation. The position that male clones, but not female ones, count as part of the endangered population arose in this context. Classifying mitochondrial DNA as biologically important was thus significant because it legitimized concerns about interspecies nuclear transfer and delimited its use.

Highlighting the biological components of cloning and chimeras served a rhetorical function. Indeed, this is a well-established aspect of classification as a social phenomenon. Derek Edwards (1991) has, for example, pointed out that classifications do not solely situate entities in particular boxes in order to create social orders. He has argued instead that classifications serve a rhetorical function and thereby allow certain actions to take place. Donna Haraway (1991: 193) has made a similar point in her discussion of positioning, in that it "implies responsibility for our enabling practices. It follows that politics and ethics ground struggles for the contests over what may count as rational knowledge." Positions on the classification of cloned "endangered" animals are forms of ethical and political action. Contestations over the classification of cloned endangered animals embody a struggle to enable certain scientific practices to take place in the zoo, which necessarily inhibits other practices in the process. These scientific practices are interlinked with certain ideas about how species preservation should be conducted into the future, and are thus world-making practices.

At the heart of the debate over cloning endangered animals are, therefore, questions over which knowledge practices should be used with endangered animals as part of conservation efforts. Addressing the new genetics generally, Sarah Franklin (2003b: 71) has argued that knowledge production is often marked by a gap between the highly technical genetic information produced in laboratories and meaningful knowledge about that information, which is necessarily socially defined, evaluated, and acted upon. Mitochondrial DNA is marked by a similar gap in zoos, where the socially produced biological facts of mitochondrial DNA must be rendered meaningful in the very specific, social situation of zoo-based, endangered species preservation. Drawing on this notion of a genetic gap, Franklin has argued that the logic of entities like chimeras is not based upon the different parts (e.g., the biological, the technological, or the social), but rather the principles that connect the different parts. Different classifications of chimeras connect nature and culture in distinct ways, which has consequences for the material-semiotics of wildlife more generally. The debates over cloning matter because different knowledge practices varyingly constitute "endangered species" and "nature" as meaningful categories.

This chapter has examined the linguistic terrain through which "cloning endangered animals" has been articulated as fraught and contested. But these classificatory practices have been worked out in both language and the everyday practices involved in cloning animals. The next six chapters of the book look at the actions that are being argued for in the different classifications laid out in this chapter. Different cloning projects articulate, in language and in practice, the meaning of nature in varied ways. Whose classificatory practices become "strategically naturalized" (Thompson [Cussins] 1998) is at stake in debates over these cloned animals, their role in zoos, and the vision of wild life that is therein enacted into the future.

2

Making Animals

What happens when cloned animals who have inherited DNA from both an endangered and a domestic animal are considered indisputably part of the endangered species? What kinds of scientific practices are enabled by this classification? What kinds of human-animal relationships are activated in the zoo through this classificatory practice? In varying ways, the cloned gaur, African wildcats, and the sand cat all embody this set of classificatory practices.

This chapter describes the scientific practices embodied by each of these animals. Gaining the technological infrastructures, tacit knowledge, and embodied skills required to do cloning and related reproductive techniques was the focus of scientists' work within each of these projects. Here scientists have sought to *make* endangered animals for the zoo. This goal has been embedded in the history of assisted reproduction within U.S. zoological parks, and has shaped the ways in which cloning has been articulated. In practice, the desire to make endangered animals has directed the kinds of animals produced by way of interspecies nuclear transfer. Symbolically, the spectacle of these cloned animals has lied in the ability to make wild life through technoscientific means. Technology is here understood as a possible solution to contemporary environmental problems that humans have created. Technology can improve the "health" of both humans and animals if research is jointly carried out. Cloned endangered animals live to represent such an imagined future.

This chapter begins by describing what cloning an endangered animal looks like within this set of classificatory practices. I focus on the microlevel practices involved at the Audubon Center for Research of Endangered Species (ACRES) in cloning an endangered sand cat. This experiment was ongoing while I was conducting the research. I then recount the mesolevel, organizational work that went into the gaur and African wild cat cloning projects based on interview material and published documents, as both these experiments had been completed when I was conducting the research. Based on the organization of cloning practices across these projects, I characterize each as experiments in technology development. The spectacle of zoo science has here lied in humans' ability to make "wild" animals in the laboratory.

Seeing Cloning

In July 2006 I made a trip to ACRES in New Orleans in order to see cloning endangered animals "in action" (Latour 1987). The ACRES laboratory is part of the Audubon Nature Institute, an umbrella organization that brings a number of urban wildlife organizations together, including the city's zoo. But whereas the zoo is located in the middle of New Orleans and is open to the public, ACRES is on 1,200 acres of gated land in the more rural West Bank. This is where the cloned African wildcats were born and live. When I visited this lab, the technical skills developed in cloning African wildcats were being used to clone an endangered sand cat.

Upon arrival at ACRES, I was given scrubs to change into in preparation for surgery, which initiated the day's somatic cell nuclear transfer activities. Entering the surgery theater with the senior scientist and veterinarian, I saw an anaesthetized, domestic cat positioned belly up, with its four legs splayed out from its body and its tongue clamped outside its mouth to avoid choking. Support staff were already in their places around the cat, which included a lab tech, an anesthesiologist, a person responsible for collecting the vials of retrieved follicles, and a person responsible for taking notes regarding the performance of the cat. In addition, another small, dimly lit room was connected to the surgery through a large, square window. This room housed a microscope used by a final person who was responsible for receiving the follicles, transferring the cells to a petri dish, and counting the number of eggs retrieved from the follicles under a microscope. I found a spot to stand next to the lab tech, where I could get a good view of the surgery as displayed on two monitors overhead.

Surgery and the day's cloning activities began. The cat's abdomen was incised. Probes were inserted to visualize the inside of the body on two computer monitors, making the ovaries visible to all of us. Forceps held the ovaries in place. The senior scientist loudly called out "Puncture. Puncture. Puncture." as the ovary was repeatedly pierced to remove the follicles. After several punctures, the ovary was wiped clean of blood and the process continued until no more follicles could be found. The process was repeated on the second ovary. A tally was made next door; sixty-seven eggs had been retrieved.

The surgery began to buzz with chatter, and everyone appeared rather happy with this outcome. The woman in charge of taking notes about the performance of the cat turned to me with a smile, exclaiming that I must be their good luck charm. I probably looked rather confused, because the lab tech continued by explaining that the domestic cats in their colony had not been producing many eggs lately. And one of their "best performers" had just done her tenth retrieval, meaning that she would be retired as a research subject and made available for adoption into a home. The lab tech said that they had sufficient eggs, and so the laboratory would be able to continue their cloning experiments over the next two days. And I would be able to continue to see how cloning was being articulated within this laboratory setting.

At this point, all the members of the team dispersed and the lab tech asked me to join her. We walked out of the surgery theater, and down the hall to the embryo lab where the embryologist, Martha, was at work. The lab tech told her the good news regarding the number of eggs, as Martha took the petri dish to view her research materials under a microscope. "Take a look, Carrie." Looking through the microscope, all I could see was tissue and blood. "I don't know what I am looking at," I said to Martha. She laughed, and said that was because she would have to clean and denude the eggs first. The lab tech said we should leave Martha to it, as we needed to return to surgery for the second egg retrieval. This preceded much like the first and resulted in about seventy more eggs for the lab to work with that day.

When the lab tech and I returned to the embryo lab, bringing the second batch of eggs to Martha, we entered a transformed space. The lights were now dimmed in the once brightly lit room and classical music played softly in the background. Martha was seated at a different microscope, with her back to us. Her work was displayed on a computer screen above her, now representing clearly discernible egg cells against a blue background that is an archetypal image of somatic cell nuclear transfer. The lab tech quietly went about arranging the second batch of ova for Martha, who called out "Hi, Carrie. Sit down."

I took a seat behind Martha, who explained her work as I watched on the computer monitor. She pulled one cell away from the group, looking for the polar body to determine if the egg was mature. If not, she would move the egg to a section of the petri dish designated for waste material. If the egg was mature, she would quickly x-ray it to see the nucleus, careful not to overexpose and damage it in the process. Martha used this image to direct her pipette through the cell wall toward the nucleus, which was then suctioned out and placed with the other waste materials. Martha then moved to the section of the petri dish that housed the somatic cells taken from an endangered sand cat, whose genome was being reproduced through the nuclear transfer process. One somatic cell was suctioned into the pipette, brought back to the enucleated egg, and inserted within the cell. After all the eggs had been re-enucleated with sand cat somatic cells and the waste material suctioned out and disposed of, the petri dish was placed in an incubator. Martha then began this rhythmic process again, which was almost lulling to watch in the darkened embryo lab in which classical music played softly in the background.

Martha got up once this next batch of somatic cell nuclear transfers was complete, placed the petri dish in the incubator, and removed another petri dish containing cells that had been micro-manipulated. She brought this petri dish to a different microscope. Below the eyepiece were probes that were brought around each cell one at a time, generating an electrical pulse that would fuse the somatic cell into the egg cell. Martha explained that the biotechnology companies often used a different machine that fused all the cells in the petri dish at once. However, given the lack of research materials available where endangered animals are concerned, ACRES opted to fuse the cells one at a time in order to increase their success rate. Fusing the cells is crucial to the somatic cell nuclear transfer process, in that it makes two separate cells into one. These cells were then put into another culture and incubated overnight. The next morning, the cells that divided properly were transferred back to the domestic cats that had acted as egg donors in the morning surgeries. These cats would serve as the surrogates to the endangered sand cat clones.

After the visual aspects of somatic cell nuclear transfer were complete, I left the embryo laboratory and shadowed a lab tech in order to watch some of the work that supports this research endeavor. I prepped to enter the domestic cat colony, watching the lab tech and animal attendant test the cats for the following week's surgeries and interspecies nuclear transfers. I chatted with the veterinarian while he took a break from painting one of the rooms in the domestic cat colony. I watched the lab tech take a sperm sample from a domestic tomcat for an in vitro fertilization experiment. The lab tech carefully recorded her activities. She mentioned that because the lab had had so many difficulties getting cloning to work, they very carefully recorded their laboratory activities so that "the system" could be constantly monitored and tinkered with.

Significantly, domestic cats were strongly present in the articulation of cloning within this lab. Domestic cats were patients in surgery. Domestic cat ova were counted, cleaned, enucleated, and transformed. The cats in the colony made up a collection of sentient beings that were fed, petted, consoled, tested, and whose living spaces required cleaning and maintenance. The policy at ACRES is that no domestic cats are euthanized, and so the lab also runs a quasi-adoption center. Meanwhile, my closest sighting of a sand cat came in the form of a somatic cell (re)imaged on a monitor. While sand cats were the object of the experiment for much of the day in the lab, the work done with sand cats was minimized and reduced to the presence of somatic cells. All other work was done with domestic cats. For scientists at ACRES, interspecies nuclear transfer and interspecies gestation were the only ways to develop a research program in cloning endangered animals.[1]

Articulating Cloning

Seeing cloning in action allowed me to understand the meaning and significance of cloned endangered animals in new ways. On the surface, the cloned gaur and African wildcats were created as models of interspecies nuclear transfer. These animals were meant to prove that it is possible to combine an enucleated egg cell from a domestic animal with the somatic cell of an endangered animal in order to create an embryo. Both ACRES and Advanced Cell Technology (ACT) wanted to make animals that would demonstrate and validate the technique. However, these laboratories also wanted to demonstrate that they themselves had the ability to do interspecies nuclear transfer, to articulate cloning as part of their everyday practices. In other words, these animals not only embodied the validity of a technology; they also embodied the technical skill of the laboratory from which they came.[2] The description of cloning practices above demonstrates this in the more microlevel practices of the ACRES lab. But this can also be seen in the mesolevel, organizational processes involved in deciding to clone the gaur, African wildcats, and sand cat to which I turn now. In each instance, the kind of animal produced through cloning reflects the focus on technology development.

Cloning a Gaur

Just two months after the birth of Dolly the Sheep, Oliver Ryder and Kurt Benirschke published a commentary in the journal *Zoo Biology* (1997). In it, they discussed the ways in which cloning could be used with existing technologies in order to do conservation in new ways. Specifically, they argued that cryopreserved somatic cells, which they had been collecting from zoo animals since the 1960s as part of their Frozen Zoo™, could be reinterpreted as protoindividuals in conservation efforts.[3] This, Ryder and Benirschke speculated, would mean that fewer living animals would be required for species preservation programs, which would free up limited space on an ever-shrinking planet. Cloning and cryopreservation could be used to preserve more species using less space, thereby addressing two key constraints in endangered species preservation efforts. But before cloning could transform species preservation in this way, the technique had to be developed and proven with endangered animals.

Meanwhile, individuals working at ACT were also becoming interested in using somatic cell nuclear transfer with endangered animals. During the late 1990s, ACT shifted its identity from a company that produces transgenic animals for pharmaceutical production to a company that creates biomedical therapeutics using human embryonic stem cells. In forming this identity, ACT found itself in the middle of politically and ethically charged debates regarding the moral status of human embryos. In response, ACT wanted to find out if embryos could be created using an egg cell donor of a different species than the somatic cell donor. Specifically, the company wanted to know if domestic cow eggs—plentiful in slaughterhouses—could be used with human somatic cells to conduct human embryonic stem cell research, and possibly create therapeutics.[4]

Scientists at ACT thought that this research trajectory could best be pursued with endangered animals, given that the company's initial transgenic work involving human bodies resulted in public concern and political protest.[5] Chief Scientific Officer Robert Lanza was aware that some people working at zoological parks were interested in using cloning for endangered animal reproduction. And ACT scientist Philip Damiani had a long-standing interest in using assisted reproductive technologies with endangered animals. Lanza and Damiani knew that endeavors to clone endangered animals would experience parallel difficulties to those experienced by human embryonic stem cell researchers: neither could allocate sufficient numbers of egg cells for their research.[6]

The first project to clone an endangered animal was born of a mutual curiosity among Kurt Benirschke of the San Diego Zoo Global and scientists at ACT in finding out whether or not it is possible to revise the somatic cell nuclear transfer process to incorporate domestic cow egg cells and endangered cow somatic cells in the cloning process. The goal was to prove that interspecies nuclear transfer is feasible, which was interconnected with both reproducing zoo animal populations in the context of less space on the planet and developing human embryonic stem cell therapies in the context of treating individual humans for various disorders. The first cloned endangered animal was thus a model of interspecies nuclear transfer.[7]

This cloning project was defined as a proof of principle exercise. The most easily available materials were used in order to maximize the likelihood of quick success. Scientists decided that the gaur was the best species to clone because there had been previous success doing interspecies gestation, wherein a Heifer cow gestated a gaur embryo and birthed a gaur newborn.[8] Previous technical success was the primary referent used to determine which species should come into relation through the interspecies nuclear transfer process. And scientists were happy to see that some of the cloned cells had cleaved into embryos, making the project a success.

The project was considered a technical success with the creation of interspecies embryos. But for the project to be seen as a success for zoos and in the popular press, an animal would need to result from the experiment. As a human embryonic stem cell company, ACT had the skill and resources needed to create cloned embryos. But they did not have the animals needed to gestate the embryos. In response, Lanza and Damiani contacted Trans Ova Genetics, a biotechnology company located in Iowa that uses assisted reproductive technologies to selectively breed cattle. Trans Ova Genetics had extensive experience doing embryo transfers as well as access to a sufficient number of domestic cows that could serve as gestational surrogates. The company agreed to participate, and transferred the cloned embryos to domestic cows for gestation as well as cared for the domestic cow surrogates throughout pregnancy. As previously mentioned, one cloned gaur was born on January 8, 2001. However, he died days later due to problems that were likely related to husbandry rather than cloning per se.

Cloning African Wildcats

While ACT and Kurt Benirschke were involved in a collaborative cloning project, Betsy Dresser and her colleagues at ACRES began to question whether they should develop somatic cell nuclear transfer to clone endangered wildlife. ACRES was founded in 1996 with the mission of developing assisted reproductive technologies for endangered species. Given that mission, the lab believed that developing interspecies nuclear transfer fell within its purview. However, the Audubon Nature Institute's leadership was concerned that this research trajectory was simply too controversial. After some negotiations, it was nonetheless decided that ACRES could initiate a cloning research program.[9]

As a laboratory premised on developing assisted reproductive technologies with endangered animals, ACRES already had many of the elements at hand that are needed to clone. This included a research colony of both African wildcats and domestic cats, with which scientists could procure somatic cells, egg cells, and gestational surrogates. The research center had already started a small collection of cryopreserved somatic cells, sperm, and embryos in their frozen zoo, offering further somatic cell samples for the cloning endeavor. Senior scientist C. Earle Pope had thirty years of experience doing oocyte retrievals, in vitro fertilization, and embryo transfers with felids and other species. And recently hired scientist Martha Gomez had experience micro-manipulating cells. As such, ACRES had a significant knowledge base, much of the technical skill, and many of the material resources needed to initiate a research program based on transposing somatic cell nuclear transfer to the zoo. In this context, they showed that zoos too could clone.

In order to move somatic cell nuclear transfer from domestic animals to endangered animals, Dresser brought additional people and knowhow associated with domestic animals to her zoological research laboratory. First, Dresser hired Philip Damiani from ACT to work at ACRES. Damiani's success in cloning the first endangered animal made him a valued addition to a laboratory that was trying to bring cloning home to the zoo. Second, Dresser and her colleagues decided that Martha Gomez should return to Australia, where she had been trained to do intra-cytoplasmic sperm injection (ICSI) with sheep, for additional training in somatic cell nuclear transfer. ICSI is a "micro-manipulation" wherein a single sperm is inserted into an oocyte using a pipette under a microscope. Micro-manipulations are time-consuming, laborious modes of laboratory work that are quite different from more surgically based assisted reproductive technologies such as egg retrieval and embryo transfer. Given Gomez's previous experience doing the micro-manipulations involved in ICSI, she was an ideal person to train in this new technique. Gomez spent one month in Australia, learning to do somatic cell nuclear transfer with domestic sheep in order to bring this set of techniques back to the zoo for use with wildcats.

In addition, the laboratory committed a significant amount of time and resources toward becoming a somatic cell nuclear transfer lab. This process required fine-tuning the practices of the laboratory in order to create optimal conditions under which each step of the nuclear transfer process could occur. The researchers in the lab had to learn how to work with each other, and the animals that they had on hand, in new ways in order to articulate somatic cell nuclear transfer into their work practices. Scientists and other laboratory workers at ACRES therefore repeatedly rejected the notion that there is anything like a "recipe" for cloning. They instead described their laboratory as an evolving "system" for making cloning work "in their hands."[10] The point was thus not only to reproduce an endangered animal using cloning; the goal was to create and sustain a laboratory system able to do this kind of technoscientific work. It was this articulation work, as much as a cloned animal, that was viewed as valuable by ACRES.

ACRES began transferring somatic cell nuclear transfer into its facilities in 2000. This research trajectory initially resulted in the birth of four different litters of cloned African wildcats.[11] Unrelated male and female cloned wildcats were bred, which resulted in two litters born on July 26, 2005 and August 2, 2005. While I was conducting this research, ACRES was using these African wildcats as models for reproducing endangered small felids. A sand cat was born in 2008, but unfortunately died about sixty days after birth. C. Earle Pople emailed me in 2010, saying that the laboratory was continuing to conduct cloning experiments but this was no longer the primary focus of their research.[12]

Illustration 2.1: Cloned African Wildcat at ACRES
(Photograph taken by Carrie Friese at the Audubon Center for Research of Endangered Species (ACRES), Audubon Nature Institute, New Orleans, La. Permission to print photograph received from the Audubon Nature Institute.)

Miles, the cloned African wildcat shown in Illustration 2.1, now stands as a witness to the scientific abilities of ACRES. He was the one living embodiment of technology development I encountered while conducting this research. His life proved that interspecies nuclear transfer is a viable mode of reproduction. Miles also demonstrated the ability of ACRES to use this technology to reproduce zoo animals. In the process, Miles had become an object in the Latourian (2004b) sense. His birth coalesced and unified the gathering of humans and nonhumans that took place within the cloning process, converting the social process of interspecies nuclear transfer into a fact. Miles has therefore lived to demonstrate that species boundaries can be crossed using interspecies nuclear transfer. But one consequence of this conversion is that what Miles has meant for the zoo has been underdetermined. He does not have any clear display value outside the research center, and cannot be seen by the general public. In addition, African wildcats are not managed by zoological parks in the United States, and so are not particularly valued by zoos. Indeed, ACRES struggled to find homes in other zoos for his offspring. Miles was thus "made to be born" (Franklin and Roberts 2006; Franklin 2006), but the life he was born to live has been limited.

Developing Technology

The experiments resulting in the gaur, African wildcats, and sand cat were organized according to the logic of technology development. The primary goal was to prove that interspecies nuclear transfer works. ACT was willing to spend $200,000 to clone the gaur because the animal represented the viability of interspecies nuclear transfer, which could help facilitate human embryonic stem cell research in both scientific and political ways.[13] Meanwhile, the African wildcats proved that ACRES was able to clone a zoo animal. The laboratory became a repository of technical skill, which could conceivably be called upon as needed in future species preservation efforts.

In order to learn if and how interspecies nuclear transfer is feasible, scientists in both instances used the cells and bodies of animals that were most readily available.[14] Philip Damiani told me that the gaur was chosen for a cloning experiment because gaur embryos had previously been successfully gestated by domestic cows. In addition, cow eggs are readily available due to the beef industry. Meanwhile, C. Earle Pope told me that the African wildcats were chosen because this species was available for them to work with. ACRES have a research colony of African wildcats. These animals are not considered endangered, and are not generally valuable to zoos. And this is precisely what makes these animals available to life science research, which is necessarily invasive and therefore physically risky. Across all these cloning projects, the focus was to prove interspecies nuclear transfer worked as quickly and as easily as possible. This focus directed the kind of animals that resulted from experimentation.

In using the cells most available to them, scientists were engaging in reproductive practices that were symbolically interlinked with the goal of increasing the number of animals in an endangered population. An article on the gaur cloning project described the significance of reproductive technologies to species preservation this way: "Recent advances in assisted reproductive techniques such as cryogenics of gametes/ embryos, artificial insemination, and embryo transfer have allowed for new methods for the further *propagation* of endangered species" (emphasis added; Lanza et al. 2000: 80). This statement emphasizes the need to "propagate" new individuals in small populations, which aligns with species preservation practices that are based on what many people I spoke with referred to as demography, or an effort to increase the number of individuals within captive, endangered populations.

It is important to point out, however, that these animals also embody different ways of doing technology development in the zoological park. The gaur was created through collaboration between a zoo and two biotechnology companies. As a human embryonic stem cell company, ACT had particular scientific and financial interests in the project, and therefore financed the research. Specifically, ACT wanted to prove the principle of interspecies nuclear transfer so that they might be able to use cow eggs, instead of human eggs, to create embryos for stem cell derivation. When conducted with humans, this experimental and controversial research was a public relations disaster. But "endangered species" tend to bring substantial support for controversial biomedical technologies, and cloning endangered animals was no different.

A scientist who worked on the gaur cloning project at ACT described the public relations benefits of cloning endangered animals to me in this way:

The endangered species came as an aside. We looked at it as a means to changing the perception people have of the cloning technology. At that time when all this work was being done, Dolly was already born but there still wasn't a general acceptance of the cloning technology. But it was really unusual when we started to work with endangered species that the perception of the technology changed completely. So it was acceptable to use this, you know, God-fearing technology for saving endangered species. But it wasn't acceptable to clone animals that we would potentially eat, or that would get into the food chain. . . . So we kind of decided to do the endangered species project to test the technology, but also to use it for PR purposes, to kind of sway people's opinion of the technology.

Interview (July 1, 2005)

God-fearing is normally used to denote a particularly religious person. In this context, it is somewhat strange to refer to cloning as "God-fearing," not only because the word does not normally describe a technique but also because this technique has been so deeply contested in religious terms. In this context, I think that the scientist is trying to elicit religious, or "God-fearing," people's opposition to cloning within this statement. At the time, this opposition was linked with the evangelical, Christian right. Their idea of embryos as equivalent to fully formed persons was the basis for then U.S. president George W. Bush outlawing the creation of new embryos for human embryonic stem cell research. ACT's endangered animal cloning project was meant to rupture this opposition. The gaur represented ACT's attempt to change people's mind about cloning, to show that it could be used for good things like preserving endangered animals. Cloning could be accepted by God-fearing people.

While ACT was interested in proving the principle of interspecies nuclear transfer in a manner that would promote the company's public image, the zoo was able to exchange cryopreserved fibroblast cells for a living, breathing animal of an endangered species. The general idea was that if cloned animals are going to be produced as proofs of a concept, they might as well serve multiple purposes over the course of their lives. In this context, a university scientist commented that the collaboration embodied by the gaur could become a standard avenue through which zoos produce animals into the future. During an interview he remarked:

"It'll be interesting to see how many of the other private companies feel like they can get some public relations benefit out of contributing to endangered species efforts" (March 14, 2006). However, as we shall see, this project raised questions about who gets to decide what kind of animal the zoo gets out of these purportedly mutually beneficial projects.[15]

Meanwhile, the African wildcat and sand cat cloning projects brought technology development into the zoo itself. This meant that the zoo had to finance this research endeavor as opposed to relying upon biotechnology companies who might want positive public relations. However, this also meant that the zoo—rather than the biotechnology company—acquired the embodied knowledge required to do cloning. I asked C. Earle Pope if he envisioned this system as providing a service function for zoos into the future. He replied that this was his dream. However, he continued to state that the focus for now had to be on learning to do these techniques, and so experimentation had to use zoo animals and zoo animal bodily parts that were most readily available. Cloning was articulated here in order to give zoos the technical knowhow required to make endangered animals when needed by zoos in the future, so that zoos would not be limited by the whims of biotechnology companies.

In turn, the zoo presumably benefited from the publicity which has routinely resulted when another animal species is cloned for the first time. Most zoo scientists I spoke with firmly believed that mass-mediated publicity of technoscientific breakthroughs in the zoo generate new forms of capital for the zoological park. Indeed, private donations were a key source of funding for ACRES at this time.[16]

Situating Technology Development in the Park

The focus on technology development—embodied by the cloned gaur, African wildcats, and sand cat—is consistent with the original formulation of the reproductive sciences within U.S. zoos. When I asked how the reproductive sciences were incorporated into zoological parks in the United States, many people I spoke with began their narratives in the late 1970s and early 1980s, shortly after Louise Brown was born by way of in vitro fertilization (IVF). At this time, three young reproductive physiologists joined different zoological parks across the United

States. David Wildt, a physiologist with a background with agricultural and companion animals, joined the National Zoo in Washington, D.C. Reproductive physiologist Barbara Durrant went with the San Diego Zoo. And Betsy Dresser, a physician, joined the Cincinnati Zoo. These scientists were reportedly excited by the possibility of reproducing endangered and other zoo animals using assisted reproductive technologies (ARTs). Wildt summarized this early enthusiasm as follows:

> We were all very young and idealistic. And I remember that we all sort of felt that there would be a quick fix. There was the birth of Louise Brown. . . . So everybody in the late 1970s, early '80s sort of thought at the same time, "Well, how can we use these high tech approaches to enhance offspring production [of wild and endangered species]?"
> *Interview (July 18, 2006)*

Many zoo scientists initially had high expectations of the reproductive technologies, which were viewed as a "magic bullet" to a new set of problems posed by rapidly rising extinction rates. At this time, the danger of many wild species going extinct had been formally recognized with the ratification of the Endangered Species Act (1973) in the United States as well as the Convention on International Trade of Endangered Species of Wild Fauna and Flora (1973) internationally. In this context, breeding increasingly became a means for zoos to sustain themselves while also assisting in species preservation. Scientists thought that technologies used to reproduce both agricultural animals and humans could be used to reproduce zoo animals within this context. The focus was on propagating wild animals for the zoo in order to increase the number of individuals in the population, in a zoo management program based in demography.

However, many wild animals are rather reticent about breeding in captivity.[17] In this context, scientists thought that assisted reproductive technologies could be used to transfer reproductive labor from zoo animal bodies to the laboratory. It was hoped that by having humans rather than the animals do the reproductive work, difficulties associated with captive breeding could be circumvented. But even if embryos could be created in the laboratory, gestational surrogates were still required for those embryos to become animals. Shulasmith Firestone's (1970) (in) famous dream of a reproductive apparatus fully removed from the body has not yet been realized in humans or animals.

In this context much of the early research done by U.S. reproductive scientists working in zoological parks was focused on determining whether domestic animals could serve as gestational surrogates for the zoo animal embryos that humans were creating in the lab. Wildt stated that reproductive scientists basically wanted to know if "you could put zebra embryos into horses and bongo embryos into cows" (Interview, July 18, 2006). Some of the original research in this area was successful and offspring were born through interspecies embryo transfer, the gaur being one example of this.[18] And the successful use of interspecies nuclear transfer received quite a lot of press. This is not surprising, given the ways in which science reporting consistently tells stories about technologies that could serve as "magic bullets" to pressing problems (Nelkin 1995). In response, Wildt found that U.S. zoos became more and more receptive to reproductive technology development in the 1980s. And this historical interpretation was reiterated in my interview with Barbara Durrant.

The incorporation of the reproductive sciences into U.S. zoological parks thus occurred considerably later when compared to academic biology, medicine, and agriculture. Adele Clarke (1998) has shown how these professions came together in disciplining the reproductive sciences in the United States, largely during the interwar years. Initially, the reproductive sciences focused on the structures and functions of reproductive organs. Clarke (1995) has defined this as "modern reproduction," wherein the focus is on understanding so as to control. She distinguishes this from "postmodern reproduction," which developed somewhat later and focuses on transforming reproductive processes. Assisted reproductive technologies are a central endeavor here. Clarke points out that modern and postmodern reproduction are neither discrete nor temporally linear, but rather deeply interrelated and often operate in tandem.

Zoological park scientists sought to get on the "bandwagon" (Fujimura 1992) of postmodern reproduction in order to transform the reproduction of zoo animals. It was generally assumed that the species intensively studied during modern reproductive science would be adequate models for zoo species. Reproductive scientists also thought that assisted reproductive technologies used with humans and domestic animals could be applied to other species rather easily. In other words, many reproductive scientists working with zoo animals hoped to bypass modern reproductive research on zoo animals and jump right into postmodern reproductive transformations.

Cloning endangered animals has thus been symbolically linked with the early development of reproductive sciences within U.S. zoos, which was focused on developing technology as a means to fix the twin problems of species extinction and wild animal collecting. In other words, these cloning projects are a site of technological optimism. Paul Wapner (2010: 86) has noted that technological optimism is generally used to *dispute* environmentalists' claims. Species extinction and global warming are not considered problems per se by these critics, but rather sites where human ingenuity and technological development are required. However, the development of assisted reproductive technologies in zoos shows how technological optimism has operated within species preservation itself. The scientists involved in cloning the gaur, African wildcats, and sand cat assumed that species extinction is a problem. But they have also believed that technology is (at least part of) the solution.

Reenvisioning the Zoological Spectacle

Science has long been a crucial component in the identity of zoological parks. In the past, the scientific identity of parks was rooted in using observation as part of natural history and ethology.[19] The cage worked to symbolically connect this scientific identity with the educational and entertainment aspirations of zoos. In this context, it is worthwhile to ask how technology development is interlinked with the educational and entertainment aspirations of contemporary zoological parks.

Seeing wild animals is a spectacular experience, or at least it is supposed to be.[20] Indeed, being able to witness wild animals, which one would otherwise never have the chance to see, has long brought people—and animals—to zoos. However, sensibilities regarding watching wildlife have been changing with the advent of film, and even more so with television.[21] The camera has become a new way to "hunt for" and "capture" wildlife, creating new ways for people to see wild animals, their habitats, and their social practices.[22] Viewing animals through film remains one of the most prominent modes of visualizing wildlife today, evidenced by the popularity of nature series like *Planet Earth* (Fothergill 2006) and more recently *Life* (Gunton 2009).

In this context it needs to be emphasized that cloned animals are *not* displayed as spectacles within the park. The cloned African wildcats are not publicly displayed at the New Orleans Zoo, but are instead housed as part of the research colony at ACRES. This means that the cloned cats are not available to the public, but are instead kept private in conjunction with the scientific practices of the zoo. Meanwhile, the cloned gaur's next of kin—the cloned banteng—is publicly displayed at the San Diego Zoo. However, the display practices work to minimize— as opposed to publicize—the fact that this animal is a clone. This was made abundantly clear to me during my first visit to the San Diego Zoo.

My partner Stephanie and I went to the San Diego Zoo during the summer of 2005 in order to see the cloned banteng. We began our visit with a bus tour of the park. I was interested in knowing what this zoo, as an institution involved in the public understanding of science, had to say about the topic of cloning in publicly presenting their cloned animal. The thirty-minute bus ride emphasized many of the animals in their collection that are often thought of in conjunction with the zoo. The polar bear, giraffes, zebras, and elephants were all discussed, and many pictures were taken. But to my surprise there was not a single word about the world's first cloned animal to be displayed in a zoo.

A bit disappointed, Stephanie and I alighted the bus. Map in hand, we continued on, making our way to the hoofstock section in order to find the cloned banteng ourselves. We carefully checked each display so as to not miss it, but eventually reached what we presumed was the end of the hoofstock section. Frustrated and rather hungry, we walked slowly on, speculating that the cloned banteng's enclosure must be one of the displays under renovation. We were discussing when we could return to the zoo, when Stephanie said that she saw what appeared to be another hoofstock display ahead. Hopeful, we walked quickly on.

Upon reaching the display, I was certain that the three animals in the enclosure had to be banteng. Based on pictures I had seen, I knew that one of the animals was the clone. Nonetheless, we began to look for the placard to verify that one of the cows in the display was indeed the product of interspecies nuclear transfer. Displays in zoos are normally clearly marked, but we struggled to find any description of the animals in this enclosure. Pushing some foliage aside, Stephanie at last found it. She called out that this was in fact the cloned banteng's enclosure. At long last we had found the animal that had brought us to the zoo, whose display disconfirmed almost all of my preconceptions regarding cloning in the zoological park.

The next day, during my visit to the zoo's research center, Oliver Ryder asked if I had seen the cloned banteng. I told him that I had, but that it was actually rather difficult to find him, as the placard was covered up by the foliage surrounding the enclosure. Ryder replied that the zoo had been very concerned about creating controversy in the park. Specifically, the mass media could have articulated the meaning of this cloned animal in negative ways. So the zoo decided to downplay the cloned animal. He said that I probably had more respect for that cloned banteng than any other zoo-goer to date.

Confirming my interpretation of the previous day's events, it was clear that the spectacle of cloning endangered animals does not operate through the medium of display within zoological parks. What we see is, after all, just a cow.[23] For the cloned animal to be a spectacle, we would need to see how it was produced. Sarah Franklin (in Franklin, Lury, and Stacey 2000) points out in her analysis of the film *Jurassic Park* that the spectacle of bringing an extinct dinosaur back to life through bioscience was displayed through its digital reproduction in the form of a movie. A film within the film accounted for how the dinosaurs were brought back from extinction through biotechnological means, which was necessary for these animals to be spectacular to both the fictional park goer within the film and to the real movie goer. With this analysis, Franklin emphasizes that biotechnologies require multiple media to be spectacular. The products of bioscience and biomedicine rarely garner fascination on their own.

Indeed, the spectacle of cloned endangered animals has to date been mediated largely in print, through the popular press, as opposed to display. When scientists involved in zoos clone an animal, the question is not so much how to display the animal in the zoological park. Rather, the question is how to communicate the birth through the popular press. And the goal of these mass mediated accounts of zoological technoscience do not appear to be solely, or even primarily, based on bringing people to the zoo. Rather, media spectacles regarding the technoscientific making of zoo animals is meant to bring in funding for the zoo, generating a new source of capital for zoological parks to pursue their scientific identity. The zoo scientists I spoke with regularly commented that people are interested in technology. Some of these people may not be members of their local zoo, but they may want to support the use of biotechnologies for species preservation. This kind of technologically based funding represented an additional revenue source for zoos, one that many people pointed out would not necessarily go to other areas of species preservation. In other words, technology can be a means of generating some people's interest in endangered species, zoos, and conservation. Indeed, Adrian Franklin (2002: 14) has noted that the modernist discourse of "mastering" and "controlling" nature has long generated *interest in* nature itself.

A number of scholars have shown that "hype" has been a crucial means for garnering funding for biotechnology companies.[24] Mass-mediated hype brings new stockholders to corporations, creating the capital needed to develop new techniques. Cloning endangered animals has been a way for biotechnology companies to get good press, which could bring in new capital through stockholders. In a similar way, the mass-mediated hype surrounding the creation of technoscientific endangered animals can bring new kinds of capital into zoos from private donors. Private benefactors have been central to companion animal cloning projects.[25] Why shouldn't private donors also be instrumental in the development of life science research in zoos? Indeed, one of the scientists at ACRES told me that private donors were among their primary revenue sources.[26]

Making Nature

I stated in chapter 1 that chimeras connect nature and culture in distinct ways, which has consequences for the material semiotics of wild life. In the context of technology development, heteroplasmic bodies containing DNA from two different species are brought together in the nuclear transfer process according to scientific priorities. The most available bodily parts are combined to create cells and animals that prove the principle of interspecies nuclear transfer and its applicability to the physiologies of endangered species. The cloned animal is thereby born of a technoscientific ethos. Its life demonstrates the scientific identity of the zoological park. Indeed, such a heteroplasmic animal could not have been made in any other way; it would not exist outside the laboratory. One of the benefits of making endangered animals through technologies like cloning is that such endeavors can bring new kinds of people to the zoo. Those interested in technology may become concerned with zoos, endangered species, and preservation through the spectacle of the cloned endangered animal. Technology can engender an interest in nature.

What are the aesthetics of a cloned endangered animal in the context of technology development? In articulating the aesthetics of American environmentalism, Paul Wapner (2010) has differentiated the sublime experience of being in nature from the sublime experience of seeing art. Wapner (2010: 69, 75) describes the idealized aesthetic of environmentalists in this way:

Many care [about the environment] simply because they enjoy the experience of visiting or immersing themselves in nature. Nature is beautiful to many people. Natural places, exotic species, dramatic landscapes, unique ecosystems, and various soundscapes provide many with a strong sense of pleasure and well-being. This aesthetic dimension involves the enjoyment, love and what some may call the soulful enrichment many people experience in nature. . . . An artificial world just doesn't sit right with many environmentalists.

Wapner contrasts this with the aesthetic experience of those who are more interested in human activity and ingenuity, which he contends is exemplified by the sublime experience of witnessing art. Wapner (2010: 100) describes this aesthetic experience as follows:

We often associate aesthetics with art. While beauty involves color, light, tone, and so on, and consists in qualities like symmetry and proportion, it is the way the artist arranges, constructs, or presents these elements that makes them beautiful. Art is, after all, a human enterprise. People envision and express themselves through art; art cannot simply be found. . . . This sense of made as opposed to found is key to the aesthetic judgment of many and hints at the difference between how environmentalists versus their critics orient themselves toward aesthetic pleasure.

Wapner argues that these two different aesthetic experiences are linked with two different ways of valuing nature, the former associated with its protectors and the latter with its detractors. Wapner contends that both these aesthetic experiences are linked with a dream rooted in foundations, the former grounded in nature and the later grounded in human culture.

In delineating these two modes through which the sublime is experienced, Wapner uses "the made" versus "the found" as a key site of differentiation. And this difference helps to explain how cloned endangered animals garner their aesthetic value through technology development, wherein human skill and the made is prioritized over an encounter. Cloned endangered animals gain aesthetic value by having been made; they are spectacles of human ingenuity. These animals are technologically created, rather than having arisen through evolution. It is the art of assisted reproduction that makes these animals interesting. Their hybrid DNA represents a kind of signature, showing that the animal was created through human intervention.

But what does making animals have to do with zoos and species preservation? Why do people want to develop these technologies with endangered species specifically? Interestingly, four of the technology developers I spoke with struggled to articulate why exactly they thought species preservation was important when questioned. But for those who could articulate why it was important, the visual experience of seeing animals that look different tended to be emphasized. For example, one zoo scientist replied: "So all these animals are going to die out. Big deal, right? But I think our grandchildren will be very unhappy if they can only see giraffes in pictures" (Interview January 8, 2006). This scientist articulated a belief that it is our responsibility to develop and use technologies so that species like giraffes can be seen and appreciated by future generations of people. In this context, changing the mitochondrial DNA of endangered animals is a possibly imperfect, but nonetheless understandable, route to follow if one wants to keep a variety of different kinds of animals on the planet, so that the sublime experience of encountering other species will be experienced by future generations.

For those centrally interested in zoos and endangered species preservation, however, such a technological approach can be unwelcome.

Technology has long been critiqued in environmental movements, particularly within the United States (Wapner 2010) but also in transnational species preservation efforts (Benson 2010: 97). This skepticism has been rooted in the belief that technologies have created so many environmental problems. In other words, technologies are responsible for "the end of nature" (McKibben [1989] 2003). Given that technologies are deemed the source of problems like global warming, many environmentalists have argued against technological solutions. This tension helps explains why cloned endangered animals are simultaneously hyped in the popular press and downplayed in the display practices of zoological parks. We tend to think of "natural kinds" (Hacking 1999) as the most real kind. An endangered animal made through biotechnology, who could only come into being through laboratory practices, may not be viewed as authentic by at least some regular visitors of zoos. Clones may not be considered acceptable replacements in reproducing the aesthetics of nature.

The dilemmas of making nature show that humans and animals are embroiled with one another in articulating cloning. Endangered species as a kind of human-animal relationship is being delineated, wherein humans must decide to use or not use available technologies to ensure other species persist into the future for human pleasure.

But in addition, humans are also being jointly worked upon through this articulation of cloning. Both humans and endangered animals are figured as in need of biological reworking, in the varying contexts of global warming, declining habitats, medical disorders, and diseases. The motifs of regeneration, through which "improvements" and "enhancements" are made, shape research that is meant to have applications for both human and nonhuman animals. Regenerative medicine and regenerative conservation have developed together, simultaneously, in the material cultures of the lab, as the bodies and bodily parts of domestic, endangered, and human animals are interchanged and transposed. The human is constituted both in pursuing certain kinds of relations with wild life and nature, but also in pursuing certain kinds of biotechnical apparatuses that can be used to regenerate human bodily parts. The next chapter looks at these transpositions with more depth, asking how technology development has enabled dual imaginaries regarding the enhancement of human and endangered animals alike.

3

Transpositions

In January 2006, I attended the annual meetings of the International Embryo Transfer Society (IETS). This is a rather small academic society, bringing together reproductive scientists who, by and large, develop technologies to reproduce nonhuman animals. There is a significant focus on agricultural species at these meetings, which is not surprising given the commercial basis for much of reproductive technology development. But there is also a committee for endangered species and companion animals within this organization, and the posters and presentations at the conference represented reproductive scientists who work with a range of different species across varying institutional contexts. As such, IETS represents an intellectual home for people who develop and use assisted reproduction technologies with and for nonhuman animals. Six of the reproductive scientists interviewed as part of this research were in attendance.

During these meetings, a paper was given detailing an experimental research program in the interspecies xenotransplantation of testes stem cells.[1] Here, Australian scientists reported that they had successfully derived stem cells from sperm of a highly valued cow breed in the global beef exchange. These stem cells were then transferred to the testicles of animals from a less valued breed of cow, which makes up a large percentage of the Australian cattle population. Scientists reported that the less valued, testes stem cell recipient produced sperm containing the genetic profile of the highly valued, stem cell donor. The scientists giving this paper speculated that they could change the population, sex ratio, and profitability of Australian cattle using this technique. They also proclaimed that this type of biological transformation in the Australian cattle population operated rather beneficially through "natural" reproduction.

I could sense a buzz in the room during this session, and suspected that this talk might be the highlight of these meetings. Indeed, after the session I heard people chatting amongst themselves about the possibilities. This excitement quickly manifested itself into new kinds of imaginaries regarding endangered animal reproduction. Two reproductive scientists spoke with me about what the xenotransplantation of testes stem cells could mean for endangered species. One speculated that testes stem cells could be transferred from endangered canines to domestic dogs, allowing common dogs to sire rare canines through "natural" rather than assisted reproduction. The reproductively prolific bodies of feral cats and dogs could be transformed into vessels for reproducing rare and endangered species.

The excitement surrounding this paper is part of a larger interest in redistributing reproductive processes across both individuals and species. This is a fascination that takes many forms across different sites, allowing for generative transfers in technological imaginations. Agriculturalists find redistributing reproduction useful because genetic inheritance can be untangled from gestation, allowing less valuable females to gestate the genomes of more desired females. Meanwhile, in zoo worlds, redistributing reproduction allows—as one scientist at the IETS meetings proclaimed—"domestics to do the reproductive work for the endangered species." Abundant animals are here called upon to reproduce the genomes of rare species.

The cloned gaur, African wildcats, and sand cat have inherited this fascination with redistributing reproduction, as part of sustaining both zoological parks and endangered species. These cloned animals not only embody a scientific ethos based on technology development, but more specifically one that seeks to bypass species boundaries in order to control and master reproduction in new ways. This chapter describes how this desire is being enacted not only in scientific practice, but also in the imaginaries of technology developers and conservationists alike. Exploring these technological imaginaries, I show how the cloned gaur, African wildcats, and sand cat have together embodied a particular set of ideas about making nature, which is embedded in a larger social phenomenon that Sarah Franklin (2006) has called "transbiology." The dream is to remake endangered species "from the inside out" (Clarke et al. 2003; McKibben [1989] 2003) so that these animals can live on and within a human-dominated planet.

Transposing Bodies and Techniques

Throughout this chapter, I use "transposition" to characterize the meaning of technology development in zoos and its consequences for our conceptualizations of nature. Here I draw upon my previous work with Adele Clarke (2012), in which we developed transposition as a concept to better understand how animal models have worked in the reproductive sciences across the twentieth and into the twenty-first centuries. In using the word transpose, we referred to both definitions given in the *Oxford American Dictionary*: 1) to cause to change places with each other, and 2) to transfer to a different place or context (Jewell and Abate 2001). As an analytic tool, we used transposition to refer to both the assertion of species similarity and to the processes whereby the bodies of knowledge and technique that come with certain infrastructural arrangements are moved to another area of interest along with the consequences that result. The concept allowed us to see some of the recalcitrant processes that delimit such transfers, which are not only biological in nature but also social and political. We concluded that the concept could be usefully extended from animal models per se to include an analysis of the practices involved in generating and working with different kinds of interspecies mixtures.

I here extend our development of transposition to ask what the transposition of domestic and endangered animal bodies means for zoos, species preservation, and notions of nature more generally. I show that the social project of transposing the bodies and techniques of domestic animals into zoo animal reproduction relies upon the presumption that nature and biology are not fixed, but are instead malleable to social designs. Ian Wilmut understood Dolly the Sheep as extraordinary because she demonstrated that there is no such thing as a biological limit, a key element in the idea of "nature" itself. Cloning had, in his words, ushered in an era of "biological control," in which the idea of a natural constraint to human activity had been surpassed (Franklin 2007b: 32). Biological control thus emphasized the ways in which biology can be made after culture.[2] Transposition, as a social process, similarly assumes that species barriers do not represent a constraint to either biotechnology or biological reproduction. By transgressing the very idea of such barriers, zoos can arguably make endangered animals in new ways. But how does this aspiration fit within the larger domain of species preservation? And what happens to our notion of "nature" and "species preservation" when it becomes something that can be made? Before addressing these questions, however, I first delineate the ways in which transposition has worked within each of these cloning projects, and the recalcitrant processes that have delimited this process along the way.

Transposing Techniques

Cloning animals of endangered species was often described to me as transferring the nuclear transfer technique as used with domestic animals to reproduce animals of endangered species. Historically, assisted reproductive technologies have been developed in the highly capitalized arenas of agriculture and biomedicine, putting zoological parks in the United States on the "receiving end" (Inhorn 2003) of reproductive technology transfers. Public funding through the U.S. Department of Agriculture, alongside the extensive capitalization of biomedical and agricultural technoscience, currently propels technological developments in and for these arenas.[3] No such funding structures are so extensively at work in conservation. Thus, since the 1980s there has been an increasing trend in the zoological community to take reproductive technologies developed elsewhere and refashion these techniques to reproduce endangered and other zoo animals. Ongoing endeavors to use artificial insemination, in vitro fertilization, and embryo transfer with endangered and zoo animals are examples of this. The uptake of somatic cell nuclear transfer with endangered animals is thus an extension of a rather long-standing reproductive project in zoos, which was discussed in chapter 1.

In a rather matter-of-fact tone, Philip Damiani—who worked on the gaur cloning project at ACT, moved to ACRES to help develop their cloning program, and was at the now-defunct companion animal cloning company Genetic Savings and Clone when I was conducting this research—described his experience and understanding of transposing somatic cell nuclear transfer in this way:

> Originally I was working for Advanced Cell Technology, which is a commercial cloning company back in Massachusetts, and their main focus

was cloning particularly cattle species. But my interests have always been in endangered species fields. So I kind of thought that, well, since we know how to clone domestic cattle, and there are some endangered species that are similar to domestic cattle, and that actually hybridize in the wild with their domestic counterparts, I thought it may be possible that we could take the tools or the techniques that we currently have right now and see if we could apply them to the endangered species. And that's kind of how it really started, was just to basically say "okay, we can clone one species, why not try another species?" . . . Because there is a lot of funding for research with livestock from the USDA, because we eat livestock, the research and the technologies used with these animals is much further ahead than what you have for endangered species. So I would say there is sort of a trickle down; it would look like a funnel.

Interview (July 1, 2005)

This statement highlights the idea that a technique is developed with a particular species, and those species are situated in certain social, political, economic, and historical milieus that have coalesced into established and institutionalized human-animal relations, such as livestock. The goal is to transfer these techniques to other species located in other types of socially mediated spaces, in this case the zoo that has fewer financial and material resources. The general idea here is that techniques should move relatively easily across these boundaries because of biological similarities.

Biological continuities across species are thus crucial for the transposition process. Indeed, the presumption of species similarities has been central to the animal model paradigm throughout much of the twentieth century.[4] Well-studied index cases (e.g., drosophila, zebra fish, xenopus levis, mice, rats, cats, pigs, chimpanzees, and humans) are used as a basis for understanding the physiologies of other species. This is possible because it is believed that species conserve biological forms and functions through evolution. As such, reproductive techniques that work with one species should work in a similar way with other, closely related species. This was why one reproductive scientist, who works with zoos and endangered species, confidently told me that any reproductive technology could work on any species if sufficient time and resources were made available for the technology transfer (Interview,

January 9, 2006). Biology is not viewed as a barrier to moving reproductive techniques across different kinds of animals; the political economies of science are.

Transposing Bodies

Somatic cell nuclear transfer has not simply been developed with
domestic animals in agriculture and then transposed to endangered
animals in zoos in a linear fashion, with a clear demarcation between
the two different species and social spaces that they occupy. Rather,
the bodies and bodily parts of domestic animals have been incorporated into the very reproduction of endangered animals. With interspecies nuclear transfer and interspecies gestation, the bodies of domestic
animals have been transposed into the very different places and spaces
of endangered and zoo animal reproduction. In other words, different
species bodies are viewed as fungible enough to be made interchangeable, not only epistemologically but also materially. Indeed, the interchangeable nature of species bodies has been central to both the "reproductive cloning" of endangered wildlife as well as "therapeutic cloning"
in human embryonic stem cell research. As such, the assumption that
species are similar makes it possible to not only transpose techniques
across species boundaries, but also to transpose different species bodies
and bodily parts into one another.

The notion that species are generally similar gains traction in zoological parks for a number of reasons. In particularly, zoo animals have
been notoriously problematic research subjects. A zoo-based, reproductive scientist I interviewed spent much time discussing these difficulties with me, which she believed are not sufficiently appreciated by
the larger scientific community.

> You need individuals to do the research with. When you're working with
> lab mice or domestic cows it's very easy to go out and get a group of
> animals to work with. Well, when you're working in a zoo, we count our-
> selves lucky if we have three or four females. Of those females, they're
> probably all involved in a captive breeding program. So, you may only
> have access to them for very short periods of time. The resources are
> incredibly limited for the researchers in the zoo community, which also

makes it extremely challenging to try and develop a technique that can become commonplace. We don't have herds of animals or a group of research animals to perfect a technique on. It's really trying to work with what you have.

Interview (April 17, 2006)

Reproductive scientists working in zoos often have an extremely limited number of animals to work with. They cannot order experimental animals, which scientists working with mice certainly can.[5] Nor is it easy to justify taking zoo animals out of breeding protocols in order to test and develop invasive, experimental assisted reproductive technologies that often have low rates of healthy pregnancies and births. Transposing domestic animals into zoo research and zoo animal reproduction is a means to work around the absence of endangered animals, which is—after all—a defining feature of this kind of animal.

The project to clone the gaur reflects this attempt to work around two different kinds of inaccessible bodies and bodily parts simultaneously through transposition. ACT wanted to work around difficult to allocate and politically contentious human and endangered animal eggs simultaneously. Here, domestic animals not only model a technical practice that could then be transferred to endangered and human bodies (e.g., interspecies nuclear transfer). Domesticated animals were also used as reproductive replacements. The logic here is that, given that the bodies of animals from different species are treated as more or less the same in the epistemic cultures of biology and medicine, then the body parts of model organisms should be more or less interchangeable with those of targeted species. Domestic animals *become* reproductive technologies in this instance.

Transposing Infrastructures

Transposing the techniques and bodies of domestic animals into zoo animal reproduction is coupled with moving the infrastructures that support domestic animals into the zoo. Adele Clarke (1987) has pointed out that the shift from descriptive approaches in natural history to experimental approaches in physiology worked to shift scientists' material needs from a limited number of specimens of a range of different species (the approach that zoological parks have been premised upon) to a large number of living but "sacrifice-able" (Lynch 1989; Birke, Arluke, and Michael 2007) animals of the same species. As a consequence of these shifting knowledge practices, a tremendous amount of knowledge has been produced regarding the biology and genomics of those animals that have been "the right tools" (Clarke and Fujimura 1992a) for the job of experimental biology. And the simultaneous commercialization of these animals has worked to solidify attention to a few select species in the life sciences (Rader 2004: 258, 260).[6] There has thus been an "infrastructuration" (Edwards and Lee 2006) of particular species as model organisms in the life sciences. Zoo animals have not been the "right tool for the job" (Clarke and Fujimura 1992a) of physiology within this context. While zoos have long defined themselves as, in part, a scientific institution, the utility of zoo animals as research objects has remained tenuous.[7]

Interspecies nuclear transfer in part garners value for those working in zoological parks because researchers could theoretically work around the lack of a biological infrastructure with endangered animals by working with domestic species. It is easier to work with domestic than wild or endangered animals for a slew of reasons, including accumulated knowledge regarding their biology, their legal status as sacrifice-able, and the easier face-to-face interactions with domestic when compared to wild animals. One of the scientists who worked on the gaur cloning project adamantly supported this, stating, "Yeah, I mean I wouldn't want to do that [cloning process] with a gaur. He'd kill me. No way. I'd much rather do that with a domestic counterpart" (Informal conversation, January 8, 2006). Because it is easier and physically safer for human and domestic animals to come into the kinds of physical contact that experimentation requires, researchers move these familiar animals into experiments on other species.[8]

Interspecies nuclear transfer can therefore be conceptualized as a means for overcoming the lack of a biologically mediated infrastructure for working with endangered animals in bioscience. Rather than work with wild and endangered species, much of the work can conceivably be done with domestic animals and domestic animal body parts. Hypothetically, one would not need to know very much about the reproductive physiology of the species being cloned because the majority of the work is done with well-understood, domestic animals. All that is needed is a cryopreserved somatic cell from an endangered animal. In other words, domestic animals do not just provide models through which researchers can learn about reproductive physiology or tinker with experimental techniques before working with rare and endangered animals. Rather, domestic animals represent an "infrastructure" (Star and Ruhleder 1996; Bowker and Star 1999) that can be deployed in order to reproduce animals of endangered species that are by definition inaccessible both in terms of their bodies and in terms of human knowledge regarding those bodies.

Disrupting Transpositions

As a practice, transposing the bodies and techniques of domestic and endangered animals emphasizes the similarities between different species. This is consistent with the development of animal models across the twentieth century, which emphasized general underlying physiological mechanisms across different species.[9] However, in practice transposing the bodies of domestic and endangered animal bodies has demonstrated recalcitrant processes of varying kinds, where biological, epistemological, and sociopolitical issues converge in resisting these assumptions of sameness.

For example, a zoo scientist told me about a failed project in interspecies embryo transfer wherein domestic surrogates were used to gestate an animal of the endangered bovine species known as anoa. The scientist commented: "I predicted that this wouldn't work because it's a totally different line of bovine and it has a very different placentation. So I, I don't think there's any hope to do that" (Interview, September 15, 2005). Indeed, managing species differences is an important aspect of transposing bodies in the process of interspecies nuclear transfer as well as in interspecies embryo transfer. For example, gaur and banteng gestate for a slightly shorter period of time than domestic cows. This difference needs to be managed in interspecies gestation. As such, basic knowledge regarding the reproductive physiology of the gaur, banteng, and African wildcats exists and contributed to the successful cloning of these animals. This contradicts the notion of domestic animals as a complete infrastructure for reproducing zoo animals.

But in addition, transposing different kinds of animal bodies has created problems for social classifications, which has legal ramifications. Through interspecies nuclear transfer, domestic and endangered species enter into a dynamic relationship and at different points in time literally change places with one another. Chapter 1 showed that classification conundrums have arisen in the official categorization of clones in zoos as a consequence of transposing endangered and domestic animal cells through the interspecies nuclear transfer process. Such conundrums have also arisen in the context of everyday practice, raising questions about how to treat the heteroplasmic individuals (e.g., as an endangered animal or as a domestic animal?).

The social process of transposing bodies requires that people engage in what Charis Thompson (1996, 2005) has called an "ontological choreography" or dynamic coordination of elements considered ontologically distinct, including domestic and wild, plentiful and endangered, as well as sacrifice-able and protected. In explicating ontological choreography as a concept, Thompson (2005) states: "What might appear to be an undifferentiated hybrid mess is actually a deftly balanced coming together of things that are generally considered parts of different ontological orders (part of nature, part of the self, part of society). These elements have to be coordinated in highly staged ways so as to get on with the task at hand" (Thompson 2005: 8). Cloning endangered animals has had to include negotiations regarding the ontological choreography of cells and animals as either "endangered" or "domestic" over the course of their development.

The importance of ontological choreography became clear to me when considering the case of the cloned gaur, which was fraught at different moments in the animal's development. Based on my conversations with people, it was clear that the fibroblast cells taken from the gaur were categorically of the "endangered species" upon leaving the Frozen Zoo™ at CRES. However, these cells became part of embryos understood as heteroplasmic, or as chimeras, through the interspecies nuclear transfer process at ACT. When the embryos were received by Trans Ova Genetics to be transferred into domestic cows for gestation, the people involved in the project had to be able to treat the embryos as more or less equivalent to domestic cow embryos in order to do the work of interspecies gestation. Oliver Ryder articulated the need for this practical equivalence as follows:

You need the infrastructure that Trans Ova Genetics has in order to do something like this. They clone transgenic cows to make pharmaceuticals. For them, there is nothing different in implanting a cloned banteng embryo into a domestic cow than there is in implanting a cow embryo into a domestic cow. Functionally there is no difference.

Interview (October 25, 2005)

Cloning an endangered animal requires an infrastructure made up of people who know how to implant embryos into an accessible population of gestational surrogates. Because this infrastructure did not exist with gaur, the cloned embryos had to be functionally the same or similar enough to domestic cows. In the context of this infrastructure, the cloned gaur embryos and fetuses had to be treated as more or less the same as domestic cow embryos and fetuses. I am not suggesting that scientists at Trans Ova Genetic thought of the cloned gaur embryos and fetuses as equivalent to the domestic cow. Rather, the scientists at Trans Ova Genetics had to be able to treat the embryos and fetuses in a manner that was more or less the same as the way they treat domestic cow embryos and fetuses in order to do the work of interspecies nuclear transfer.[10]

The ontological choreography of the cloned gaur posed two critical problems in this context. First, researchers from ACT sacrificed three fetuses resulting from the cloned embryos in order to assess for normal development and the genetic origins of the clones. This is normal practice within the life sciences, and is considered perfectly acceptable with laboratory and domesticated animals. However, after announcing this practice in a scientific journal article,[11] some conservationists argued that these sacrifices violated the Endangered Species Act. The ESA (1973) states that no animal of an endangered species, nor part of such an animal, can be harmed in any way. Philip Damiani told me in an interview (July 1, 2005) that if the gaur had not been reclassified from endangered to threatened, the entire project would have been found in violation of the Endangered Species Act.

Second, within the ontological choreography of the gaur, birth represented an important turning point. When the fetus became a newborn, the animal could no longer be treated like the domestic cows at Trans Ova Genetics but instead had to be treated like a gaur. It was assumed that the domestic cow would not rear her now alien, endangered off-spring. This meant that the researchers from Trans Ova Genetics and ACT—who together had extensive experience working with domestic cows but little to no experience working with gaur—found themselves in a position where they had to rear the newborn. Hand-rearing newborn zoo animals is generally avoided in zoos, as having the biological mother rear the infant is favored. As such, there is often scant knowledge available regarding how to rear infant zoo animals. In this context, Mike West told me that scientists from both Trans Ova Genetics and ACT held an emergency conference call to deliberate on how to rear the clone (Interview, July 18, 2006). It was decided to feed the gaur milk that appears to have given the newborn dysentery, which caused him to die just days after birth. One of the deadly ironies of developing interspecies nuclear transfer has been that, when cloned endangered animals have been born, seemingly against all odds, basic husbandry problems have frequently brought about the untimely death of these animals.[12] The question of how to care for the cloned animal has all too often been marginalized in technology development.

The classificatory questions raised by transposing bodies of knowledge and technique matter not only for the official classification of endangered species. These questions also shape day-to-day interactional practices with other species that are central to life science research, and can have life and death consequences. Managing the ontological choreography of hybrid entities like cloned, heteroplasmic animals is required for both "biological" and "social" success. These animals must be treated as more or less the same as domestic animals in certain instances for the process to work. But these animals must also be treated as different in other instances. Delineating the ontological choreography of such hybrid animals has been a crucial, if underestimated, component of making cloning work.

Transbiology in the Zoo

Transposing bodies of knowledge and technique is bound up in a more general move toward what Sarah Franklin (2006) has called "transbiology." Franklin (2006: 171) defines transbiology as "a biology that is not only born and bred, or born and made, but *made and born*" (emphasis in original). Here Franklin emphasizes that nature and culture get inverted in the working practices of an embryo laboratory, in a manner that puts into question the notion that culture comes after nature.[13] As part of this conceptualization, Franklin draws on Donna Haraway's notion of "trans," with which she describes "the shape-shifting categories by which new hybrid entities, such as transgenic mice, 'blast widely understood notions of natural limit' or kind" (Haraway in Franklin 2006: 170–171). Franklin uses "trans" to articulate the new kinds of mixtures that are being actively made through the life sciences, mixtures that would not otherwise be born.[14]

Transposing the bodies and techniques of domestic animals into endangered animal reproduction is embedded within this move toward the transbiological that Franklin describes. Endangered species are "made to be born." Cloning and the assisted reproductive technologies are premised upon making wild animals in the lab in order to work around the variety of factors that limit their natural reproduction both in and ex situ. These endangered animals are made by mixing things that are generally thought to be foundationally different, creating "unexpected and apparently unsuitable unions" (Franklin 2006: 176). This includes not only the bodies of domestic and wild species, but also the social spaces of farms and zoos, the institutions of agriculture and species preservation, and the practices of slaughter and conservation.

Franklin developed transbiology through her empirical investigations of the interface between assisted reproduction and regenerative science. In this context, Franklin (2006: 176) has contended:

> [T]he transbiological is not just about new mixtures, playful recombinations of parts or new assemblages: it is fundamentally defined by the effort to differentiate these dirty descent lines into functional, safe and marketable human biology . . . New parts made from regenerative technologies must be standardized, regulated, ethically sourced and validated before they can become part of our biology.

As such, Franklin counters the prevalent notion that life scientists are making unlikely mixtures between humans and animals, clinic and laboratory simply because they can. She counters that these mixtures are being purposefully made in order to develop a new kind of biomedical regime based on regeneration, which has consequences for economic development and health economics, individual and population health, as well as legal and clinical modes of governance.

Here the articulations of cloning embodied by the gaur, African wildcats, and sand cat diverge from Franklin's description of the transbiological. These cloned animals were made as proofs of principle. The goal was to prove that interspecies nuclear transfer was possible. This does raise a question that is important to many people: were these animals created simply as new mixtures, rather playfully showing that different bodily parts of different species can be recombined? The gaur, African wildcats, and sand cat showed how the infrastructures of domestication can be brought to the zoo, to pursue the scientific identity of the park. But how do the resulting cloned animals contribute to social projects under way in zoological parks? How does cloning assist in species preservation? The articulation of cloning across these three projects has not answered such questions.

This was a core criticism that I repeatedly encountered while conducting this research. Here are some examples of how these concerns were articulated during interviews, often with a fair amount of frustration and even anger:

A lot of the initial interest is in just showing the feasibility of it. You hear a lot about "world's first" this or that. And it drives us crazy because to me that's the easiest part of anything. You do enough procedures and if that animal, if it's physiologically similar to something that's worked before, you do enough procedures and over time eventually something will work. And you point to the one offspring and say "we've done it" and then you walk away and nothing else ever gets done with that. And to me the feasibility is the easiest part. It's making it efficient enough that you can consistently produce offspring that way [that is difficult].
Interview, Zoo scientist (April 8, 2006)

Another problem that I had with the gaur [cloning project] was that these animals reproduce extremely well in captivity. Gaur are practically a dime a dozen in captivity and reproduce—in every zoo they live in they reproduce extremely well. So that was merely a demonstration of technology, which I did not think should happen with an endangered species. . . . I have a personal philosophical problem with doing something just because you can or because it might attract attention, especially in the press.

Interview, Zoo scientist (May 16, 2006)

Knowing a lot of these individuals—I don't think they're sinister; I don't think they're Frankenstein creators. I think they simply haven't thought about this. They're so gung-ho on the technology that they think "here is a new solution that is going to prevent the death of endangered species and we've made a wonderful creation." True, but in context. Like any technology it has to be seen in context. And it has to be arbitrated and mediated in context.

Interview, Field conservationist (April 26, 2006)

Taken together, these statements critique technology development for an excessive focus on demonstration projects that fail to address the question of why and how such techniques and resulting animals could be useful to zoos and species preservation. In the first two statements, different scientists argue that there is a gap between technological development and the ways in which zoos routinely go about reproducing their animal populations. The first statement contends that insufficient attention is paid to making a technique efficient enough to become a standard tool in zoo animal reproduction. The second statement articulates the contention that technology developers focus on the wrong species whose reproduction isn't even in need of technological intervention. As a consequence, little attention is given to the zoological or conservation context in which these technologies must necessarily enter in order to be made meaningful. The third statement categorizes this situation as an instance of technophilia, wherein the potentialities of technology are pursed as an end in itself rather than in the context of its use.

I asked people involved in technology development what their response was to such criticisms. They would commonly respond that technology development is a slow process. What mattered right now was not necessarily the application of the technique, but rather learning to engage in such technical processes. In this context, it was important to bring techniques like somatic cell nuclear transfer home to zoos in order to learn how to engage with endangered animals through biotechnologies. Biotechnologies may be a crucial "tool" in saving endangered species on the brink of extinction in the future. In this context, technology developers would respond that it is crucial for zoos to develop the embodied skills involved in using these techniques. The utility of tools like the assisted reproductive technologies generally and cloning specifically was thus put off for a future time. In the process, many disengaged from their critics.

People working both inside and outside zoos commonly described conservation as a "toolbox" that contained many different tools, including habitat preservation, education, genetic management of populations, and—according to some—assisted reproductive technologies. I am somewhat skeptical of the "tool" metaphor.[15] My concern is that it leaves open the question of what role tools like cloning will play in zoos and conservation, both in the present and in the future. In addition, the metaphor risks representing the different tools available as equivalent and neutral. The tools metaphor tells us nothing about how different tools can be brought together to do particular jobs, whether or not these are the best tools for the job at hand, and who gets to decide which tools are right.[16]

Decades of technology studies have taught us that tools, while not deterministic, nonetheless "have politics" (Winner 1980). The ways technologies are created matters. A set of assumptions regarding the nature of the problem at hand, and how that problem should be addressed, are embodied by the technology itself.[17] In other words, technologies are not inert but instead have "agency" (Latour 1999). In this context, I think it is crucial to ask what assumptions regarding nature are embodied in transbiological approaches to wildlife conservation. What kinds of potent imaginations are enabled by technology development rooted in transposing bodies and techniques? As a field conservationist pointed out to me, these speculative futures may represent a whole new way of thinking about what nature is and how we should go about preserving it. And it is for this reason that we cannot think of cloning—or any other technology of preservation for that matter—as either neutral or deterministic. Technology development pursued through the logic of transposition emphasizes a technological solution to species endangerment, which seeks to change and control the biological and social reproduction of species.

After Culture: Reimagining Wild Life

What does transbiology do to the zoo? In the context of human embry-onic stem cell therapeutics, Franklin (2006) has emphasized that het-erogeneous assemblages and unlikely mixtures are being created in order to make new kinds of medical objects that can travel through existing regulatory regimes informed by pharmaceutical development and distribution. Meanwhile, the heterogeneous assemblages associ-ated with transbiology in zoos are pursued to re-create already existing endangered animals. Whereas the transbiological is about creating *new* kinds of additive human biologicals in biomedicine, the transbiologi-cal is about re-creating *old* kinds of organisms in species preservation. What difference does this difference make?

I unpack what transbiology does to our notion of wildlife and wild life in order to consider how these developments open up new ways, and reproduce old ways, of thinking about nature more generally. Early in the attempt to theorize assisted reproductive technologies, Marilyn Strathern (1992a: 22) pointed out that "the more facilitation is given to the biological reproduction of human persons, the harder it is to think of a domain of natural facts independent of social intervention." A number of scholars have similarly argued that biotechnologies are challenging the notion that society comes after nature.[18] Further, Strathern argued that this shift has consequences for the way people think about their relations with one another. Drawing on this field of scholarship, I ask here what it means to make wild life that comes after culture and society.

Preservation practices are generally premised upon the notion that nature comes before culture. For example, Stephanie S. Turner (2008: 60) has noted that wilderness preservation is based in the European romantic idealization of the natural order as divine and human experi-ence of that order as transcendent.[19] She states that while conservation to some extent requires domesticating the natural, most conservation-ists have done so in order to reestablish natural orders that existed prior to human intervention. "Conservation and preservation efforts are, in short, efforts to make right, through new technologies, the unintended wrongs resulting from older technologies" (Turner 2008: 61). This is precisely the logic we saw at work in the decision to hybridize the Florida panther and the Texas puma discussed in chapter 1, wherein a prior natural ordering justified a parallel bio-social-technical ordering in contemporary preservation practice.

Transposing the DNA of endangered animals into the cells of domestic animals poses a problem in this context. Natural orders are not being reestablished here. Rather, new kinds of organisms are being created in the laboratory, organisms that could only come into being within the laboratory itself. This difference was at times erased through the discourses researchers used, which worked to naturalize transbiology and situate it in mainstream preservation practices. However, this difference was also at times taken up to think about new ways in which preservation could be done through transbiological approaches.

Some scientists I spoke with tried to erase and minimize the significance of new genetic relations produced in the lab. This erasure was often accomplished rhetorically by naturalizing biotechnology. Here it was argued that humans have always been part of the making and remaking of the natural world. Wheat was commonly used as an example to argue that separating out human activity from the natural world simply did not make sense. In this context, people would ask why humans shouldn't use their technical skill to help keep species on the planet. After all, they would comment, preserving endangered species is the right thing to do, with whatever means possible, in the context of rampant extinction. A university scientist I spoke with, whose career has been focused on developing biotechnologies opposed to doing conservation, articulated this position as follows:

> My thought is that it is just causing evolution to happen a little bit more rapidly. We're causing, we're part of the problem, the human species, so we ought to use our intellect to try to help them out with that, with the changing world we are imposing upon them.
> *Interview (January 9, 2006)*

Turner (2008: 56) argues that statements like this challenge "the 'naturalness' of natural history, suggesting that human beings have played a key role in writing it and, more significantly, urging that we revise it through whatever means we have available, be it international treaty or interspecies embryo transfer." I agree. But I would add that statements like this also intersect with a more general set of changes that have been occurring amongst habitat conservationists since the 1990s. Conservationists focusing on ecosystems as opposed to individual species have similarly argued that humans cannot be viewed as separate from the natural world. Rather, these conservationists contend that the role of humans in ecosystems has been sustained over time, and must be accounted for and explicitly addressed in order to keep these habitats into the future.[20] Ideas about nature are changing within environmentalism more generally, wherein the separation between humans and animals is being problematized and a foundational approach to nature brought into question.[21] The naturalness of natural history is thus being challenged in different ways, and with different consequences.

Naturalizing the transbiological required ignoring the novelty of genetic relations being produced between domestic and endangered animals today. However, other people I spoke with would alternatively argue that the new genetic relations produced in the laboratory could provide a new kind of socionatural order, one that would rather significantly change the meaning of nature. As previously discussed, disorderly entities and relations can be ignored or abolished in favor of sustaining the status quo. But, as Mary Douglas ([1966] 2005) has emphasized, disorderly entities can also be embraced in order to create new kinds of orders and relations.

Genetic engineering combined with cloning formed a potent site through which speculative futures regarding the future of conservation were articulated in the context of transbiology. The field conservationist I spoke with speculated that this technological combination could be used to change the behavior of wild animals, so as to protect humans from harm. He articulated this possibility for preservationists as follows:

> What ultimately matters to us are our children. And if the big dramatic species like elephants and lions and so on pose a threat to our kids there's going to be a whole lot of people who say, "Well, why should we have species which threaten our children? Why don't we just tweak their genes and have elephants which are a lot more friendly towards people? Why don't we just tweak the genes slightly in a lion and have lions which might kill antelopes but actually are predisposed not to kill people?" Those are the sorts of things I think we're going to be up against and they're very hard moral choices.
>
> *Interview (April 25, 2006)*

Habitat conservationists and species preservationists alike are continually hindered by the fact that some of the wildlife that they are working to preserve will prey on livestock, and will occasionally kill the people who care for livestock. In this context, should humans strategically make and remake animals to fit the human-dominated world that most species must now live within? Should lions and elephants be changed genetically so that they won't hurt people, in order to facilitate their conservation? Where James Lovelock (in Clark 1997: 90) has imagined using genetic engineering to terraform other planets, the conservationist here imagines genetic engineering to terraform this planet by reworking wild animals from the inside out. Where the critics of cloning may fear the kinds of pathologies unleashed by such hybridities, here we see the idea that promoting life through hybridity could be a way to resurrect a failing nature.[22]

This scenario is far-fetched and certainly does not represent technological capabilities. However, another zoo scientist articulated a similar theme in a conversation with me, but in a manner more closely linked with contemporary technical ability. He told me that a number of bird species were becoming threatened as a result of avian malaria. In this context, he wondered if genetically modifying birds so that they were disease resistant could be a potential solution. Like the scenario above, the idea here was that many species are not well equipped to adapt to a planet dominated by humans. In this context, should available resources be used to remake the genomes of endangered species, so that they can better survive these interactions?

These visions intersect with more general speculations regard genetic engineering, genetic modification, and biodiversity. Freeman Dyson (2006) has written an article in which he envisioned a world in which biotechnology was no longer centrally organized, but instead dispersed in a manner that parallels computers. In his imagined future, amateur breeders could make new varieties of hybrid fauna, propelling a kind of domestication based on diversification rather than the proliferation of sameness currently seen in corporate agriculture. Dyson anticipated that this would represent a better future: "We are moving rapidly into the post-Darwinian era, when species will no longer exist, and the rules of 'open source' sharing will be extended from the exchange of software to the exchange of genes. Then the evolution of life will once again be communal, as it was in the good old days before separate species and intellectual property were invented." For Dyson, species boundaries are a problem that should be overcome in a free market exchange of genetic information. This would create more diversity than the planet has ever seen before.

The conservationist I spoke with used this article to argue that there may be new ways of doing conservation into the future, based not only on reconstructing ecologies but also wild animal bodies.[23] This was understood to open up a whole new way of thinking about biodiversity amongst conservationists, wherein biodiversity could be actively made as opposed to preserved through human intervention. He commented:

> We now have the technology to create any hybrid that we care [to create] and we have the technology potentially on the horizon to create diversity like the world has never seen. So if we want diversity, let's create it genetically. If we want to have animals fulfill a particular ecological niche, which is emerging and which was not there before, we potentially have the genetic and the technical skills shortly coming up to be able to accelerate the rate at which species will fill newly emergent niches in the world that we create, the human dominated niches. That's a whole new landscape. So we have some very hard decisions to make.
>
> *Interview (April 25, 2006)*

Wild Life: Endangered Species 2.0

Can we still use the term "wildlife" to denote animals that may not be tamed, but have nonetheless been remade from the inside out by humans in order to fit the planet as it exists today? Or do we need another term? Does the dream of a genetically modified endangered species represent a new kind of animal? Might we refer to this new landscape as endangered species 2.0, which stands for one variety of wild life?[24]

In chapter 2 I noted that technology development conjoins humans and animals through not only the elaboration of "endangered species" as a type of human-animal relationship, but also through the material practices of the lab. Interspecies nuclear transfer was being developed to jointly work on and improve humans and endangered animals through a regenerative medicine that seeks to improve the health of human individuals and endangered populations. In turn, the fantasies of remaking endangered species to fit a human-dominated world are embroiled with the fantasies of transhumanists, who are hopeful that the human-technology interface will radically extend, improve, and enhance human life.[25] Just as transhumanists desire a future in which humans can control their own evolution through biotechnological means, one of the dreams underlying technology development in the zoo has been to control the evolution of endangered species. If the human-technology interface has created a planet on which wild animals can no longer live, and humans want to continue to see these animals, why not rebuild animals at the molecular level so that they might be able to continue to live on *this* planet, as it has become? Indeed, one technology developer commented to me that it would be unethical not to use cloning in order to save endangered wildlife.[26] Such statements are conjoined with arguments that it is unethical for people not to enhance themselves using biotechnologies. Both are firmly rooted in finding technological solutions for problems, and are thus embroiled in a long history of believing that human mastery is the basis for progress.

With the focus on using technologies to biologically engineer new kinds of improved life forms, endangered species 2.0 may appear, at first glance, clearly interlinked with Clark's (1997, 1999) conceptualization of wild life. Clark uses the imagined terraforming of other planets, computer programs through which life is redesigned, and science fiction tales such as *Jurassic Park* and *Permutation City* as examples through which he delineates this concept. However, transposing the bodies of endangered species and the corresponding imaginaries of endangered species 2.0 also diverge from Clark's delineation of wild life in crucial ways. For Clark, wild life denotes a moment in which the unintended consequences and corresponding uncertainties that have delimited the modernist impulse to control, order, and master nature are not looked upon with horror and fear, but rather with intrigue and fascination. For Clark, wild life embodies the desire to facilitate an experimentation in which the resulting products exceed the specifications of designers (Clark 1999: 149). While Dyson does appreciate the productive unruliness of a cultured nature released from human control that is central to Clark's delineation of wild life, endangered species 2.0 does not explicitly incorporate these uncertainties into its experimentation. Transbiology in the zoo has not been rooted in being surprised by how biological organisms act back in innovative ways. The radical act of transposing species bodies with interspecies nuclear transfer reproduces rather than transforms the long-standing and rather conservative discourse of human mastery and control. Like transhumanism, endangered species 2.0 fails to engage in the multiple kinds of social and biological recalcitrances that have delimited the modernist dream of improvement and mastery.

Transhumanists have been consistently silent on the issue of power, refusing to consider how their fantasies of biological control reproduce corresponding hierarchies that developed in modernity, including gender, sexuality, race, ethnicity, class, nation, citizenship, age, and (dis)ability.[27] The trope of control combined with an abnegation of politics risks reproducing inequities in conservation as well. The technology developers I spoke with were often hesitant about engaging in the politics of their research. The metaphor of the tool allowed them to bypass their critics. And this hesitation about engaging with critics may actually result in another site of recalcitrance to their work.

Many people told me that the conservation community would be resistant to anything like the endangered species 2.0 presented here. Such a scenario represents nature as something that humans make and control rather than something that exceeds human knowledge and intervention. Endangered species 2.0 is not created through the dynamics of intended and unintended consequences that arise through relations within and between species. This is a wild life created through human selection, wherein the life sciences are the source of awe. This vision asks other species to adapt to a set of human concerns, rather than asking humans to change their practices in order to ensure that other species have space on the planet to evolve in ways that are less directly managed. Endangered species 2.0 embodies the "dream of mastery" (Wapner 2010) that much of the environmental movement has fought against. When I asked the field conservationist I had interviewed about where he would like to see conservation go into the future, he commented:

My inclination—perhaps because I'm an old-fashioned conservationist in that sense—I would like to think that in the process of creating a more human-dominated world, we can find space within that world to accommodate other species and to give them space, if you like, to follow their own evolutionary pathway.

Interview (April 25, 2006)

Endangered species 2.0 is only one way to make nature into the future. There are alternative ways of hybridizing nature and culture in the laboratory, which reject the notion of human mastery seen in the imaginaries that have been enabled by transpositions. The problem with mastery as a metadiscourse is that it forgets all the recalcitrances that delimit such efforts in the nitty-gritty practices of scientific work, where the social and biological, natural and cultural interconnect. I will now turn to an alternative articulation of cloning, one that uses cloning as part of an effort to shepherd rather than master endangered animals. This articulation does not bypass the politics of cloning endangered animals, but instead attempts to actively address and engage in such debates while carving out a different kind of wild life.

4

Reproducing Populations

This chapter explores the scientific practices that are argued for when male animals produced through interspecies nuclear transfer are classified as part of the endangered species population, but female clones are not. The cloned banteng embodies this set of classificatory practices. While this animal may be like a hybrid according to some people, he has nonetheless been strategically included within the North American Species Survival Plan for the banteng. On what basis does it make sense to include heteroplasmic male, but not female, endangered animals within such management protocols? How are nature and culture connected in this set of classificatory practices? What visions for the future of species preservation are enabled by the scientific practices that result?

The classificatory practices embodied by the cloned banteng enact a pragmatic and pluralist approach to technology development. Questions about the salience of cloning for zoos are not deferred to the future; experimentation is instead explicitly organized with the protocols that zoos routinely use. Rather than remake endangered animals to fit a human-dominated world, cloning is instead imagined as a way to reproduce endangered populations in new ways. The focus is less on changing the genetics and physiologies of endangered animals, and more on circumventing existing spatial orders so that zoos and species preservation can be done in more efficient ways. In this context, scientists do not necessarily want to bypass and transgress biological limits and species boundaries. Rather, the biological— specifically the genetic—is used as a guide to delimit biotechnology development. As a result, the cloned banteng does not embody the science of cloning per se; he instead embodies the science of cryo-preservation and the practices genetic management. These practices represent a way to sustain the genetic nature of the planet through human management and technology, wherein the goal is to preserve as much of that nature as it exists today.

This chapter begins by describing what cloning an endangered animal looks like within this set of classificatory practices. I begin by describing the microlevel practices that I witnessed at the Frozen Zoo™, which were crucial to the mesolevel organizational practices involved in cloning the banteng. Based on these practices, I characterize the banteng as an experiment in pluralist technology development. The spectacle of zoo science has relied on humans' ability to remake endangered populations in new and more efficient ways. Specifically, cells and fully formed animals can be acted upon as, more or less, the same.

Seeing Cloning within the Frozen Zoo

I visited Conservation and Research for Endangered Species (CRES), the San Diego Zoo Global's research center, in the summer of 2005. Thirty minutes north of San Diego, the research center is located next to the Wild Animal Park in the agricultural town of Escondido. The Wild Animal Park is also managed by the San Diego Zoo Global, and offers a safari-inspired tour of animals that live on larger tracks with other species. But where this park is open to the public, the fenced and gated veterinary hospital and research center located next door are off limits. In order to gain entree into these facilities, I had to call my host.

I had arranged to visit CRES in order to interview a scientist who was involved in the banteng cloning project. I was also promised a tour of the research facilities. Unlike ACRES, this center has many different divisions that are all devoted to conducting life science research with wild and endangered animals. Indeed, my visit was with the Genetics Division as opposed to the Reproductive Sciences division. That said, like my visit to ACRES, I had come to San Diego and to CRES in order to see how cloning had settled down within this particular zoological organization.

What I saw that day bore little resemblance to what I had seen at ACRES, or what would normally be designated as "cloning." I did not see interspecies nuclear transfer being conducted, nor did I see any cells or surrogate animals. Indeed, interspecies nuclear transfer had never been done in these facilities. Rather, much of the tour was devoted to showing me the Frozen Zoo™.

The Frozen Zoo™ is a collection of cryopreserved cells housed in three large, round, industrial-looking freezers within a small, locked office of the Genetics Division. The room was stark and white, with the only zoo-like image being a colorful sign posted on the back wall that said "FROZEN ZOO." I was told that the freezers contained cultured fibroblast cells, gametes, and embryos from over 4,300 wild species of mammals, reptiles, and birds. Attached to the office door were twenty to thirty laminated pieces of paper, each of which detailed the species relations of recently acquired cells. My tour guide quickly flipped through the pages to show me some of their newest "animals," but paused at the picture of a bird. Quietly and respectfully she stated that they had recently received a cell sample from what may have been the last living animal of this species.

I was told that the Frozen Zoo™ has developed very formal policies for utilizing this collection. Every cell sample is propagated to eight vials upon receipt, with four vials stored here in the Genetics Division and the other four at an undisclosed location. The collection is thus protected from an attack or accident that could destroy its contents. At least two of the vials must be stored in perpetuity, and so are not available for either research or regeneration. If the number of vials for an individual animal is down to four and there is a request, the vial is first propagated so that one can be frozen and the other made available for research or reproduction. Four vials for each individual should thus be stored in perpetuity, with two never being thawed.

Next to the Frozen Zoo™ was a cell culture lab, containing hooded benches for culturing cells when new donations arrived. Unfortunately, there weren't any new acquisitions that day and so the cell culture lab was empty. Rather, much of the action was taking place in a darkened office next door. Here a researcher and postdoc were karyotyping an anoa, another endangered bovine species that is native to Southeast Asia. Karyotype refers to the number and appearance of chromosomes in the nucleus of a cell. When karyotyping, pictures of the chromosomes are displayed in matched pairs according to size and position. In zoos, this technique has been used to assess for infertility and to determine if zoo animals are actually hybrids.[1] A wall of this room was lined with binders, each containing the karyotype of each cell cryopreserved in the Frozen Zoo™.

As the researchers worked at the microscope, they told me that the computer program they were using to do this karyotype had been recently acquired and represented a major advancement in their work. Karyotyping is a rather long-standing practice within genetics, and so I was surprised to hear this. They continued to explain that this computer program had previously assumed 46 chromosomes, making the technology useful for work on human genetics. But anoa have 48 chromosomes. The researcher pulled a binder down from a wall to show me what it used to be like to karyotype by hand.[2] These researchers commented that they relied heavily on all the technology produced for studying humans as part of biomedicine, but that the technologies rarely transferred smoothly to zoo species. In other words, they were well versed in the multiple kinds of recalcitrant processes that delimit a smooth transposition of techniques into the zoo.

After the karyotype of the anoa was complete, my tour guide continued to show me through the different divisions of CRES. In one division, I was amusedly shown a poster presentation, which included the picture of a man ejaculating an endangered horse to get a sperm sample. In another division, a scientist showed me a slew of vials containing elephant feces awaiting analysis. Across this tour, it became clear that the life science research program at CRES was far more diverse than what I had seen at ACRES. And where interspecies nuclear transfer brought ACRES together, this technique was marginal to the everyday work of both the Genetics Division and CRES as a whole.

Learning to clone was not the key goal of any scientist I met at CRES. Rather, cloning represented an opportunity to use the somatic cells that had been cyropreserved in their Frozen Zoo™ in new ways. The cloned banteng was created in order to explicitly demonstrate the reproductive potential of cryopreserved somatic cells for the ex situ preservation practices of zoos. This focus meant that articulating cloning within a zoo lab was not the central task. Rather, these scientists wanted to articulate cloning with both their Frozen Zoo™ and Species Survival Plans (SSPs). CRES scientists were thus interested in developing cloning for the zoo.

Articulating Cloning

As previously mentioned, the cloned gaur, African wildcats, and sand cat have all sparked tremendous debate within and across zoos, which was a recurrent theme in my research. People would often use interviews with me to argue that cloning is expensive, inefficient, and invasive. Trying to develop this set of techniques is premature, particularly given that artificial insemination—the most basic of assisted reproductive techniques—remains elusive in most endangered species. Other people I spoke with were concerned about the publicity surrounding cloned animals, which made biotechnologies appear like "magic bullets" for species extinction without referencing their extremely limited use. Overenthusiasm for a technological fix could erode public support for more traditional and proven preservation practices. And most people struggled to see how these cloned animals counted as species preservation. Gaur do not require invasive reproductive techniques to reproduce in captivity. African wildcats are not endangered, and the Taxonomic Advisory Group for felines recommends that zoos not keep this species in order to make more space for endangered cats. As such, these cats were actually a problem as far as population managers were concerned.

Similar criticisms of cloning endangered animals were raised within the San Diego Zoo Global immediately after the birth and death of the cloned gaur.[3] But in addition, two people I spoke with, who worked at CRES during this time, also voiced concern that the gaur cloning project was conducted without Society-wide consultation. They believed that this resulted in a project that prioritized the needs of biotechnology companies and thus was not truly of mutual interest. While Society-wide involvement may have slowed the collaboration down, consultation would have ensured that the cloned animal would suit the needs of the zoo community.[4]

After the birth of the gaur, both ACT and Trans Ova Genetics wanted to conduct another cloning project with an endangered animal. However, extensive discussions were organized within CRES. The Society asked itself whether or not they should engage in another cloning experiment. And if they were to collaborate once again, what kind of animal might the zoo want from such a project? The criticisms of the gaur cloning project were thus publicly articulated and engaged with.

In this context, the Society's board asked Oliver Ryder, head of the Genetics Division, to write a white paper detailing when the zoo should and should not involve itself in cloning experiments. Ryder used the white paper as an opportunity to enact an alternative vision of cloning. He determined three conditions that had to be met for endangered animal cloning experiments to go forth with San Diego Zoo Global.[5] The first condition was sufficient knowledge about the species being worked with, including not only reproductive biology but also husbandry. With this condition Ryder addressed the husbandry problems that arose in caring for the cloned gaur upon birth. He also marginalized the goal of redistributing zoo animal reproduction across species boundaries in order to work around the lack of basic knowledge regarding the physiologies of these species. The second condition was the availability of a somatic cell from an endangered animal that embodies genetic information not represented within the captive population. Here, Ryder created a strong relationship between cloning and the Frozen Zoo™. Specifically, he showed how cryopreserved somatic cells could be regenerated to form a living animal that embodies genetic information lost from the captive population. The third condition was the existence of a Species Survival Plan (SSP) for the animal being cloned. SSPs manage the gene pool of captive populations that are endangered. They selectively breed animals to produce genetic diversity so that the captive population is sustainable. As such, a cloned animal embodying genetic information that is not represented within the captive population would appeal to this group. Through SSPs, the cloned animal would matter for the ex situ preservation work that zoos routinely engage in.

During an interview, Ryder showed me the ways in which he conceptualized these conditions, as a Venn diagram made up of three overlapping circles. Through the Venn diagram, Ryder delimited the species that were candidates for cloning experiments. And within the middle of that Venn diagram, Ryder commented that only males were viable candidates. This caveat was necessary to address the concern that mitochondrial DNA from a domestic animal was entering the endangered population through the interspecies nuclear transfer process. It is generally believed that mitochondrial DNA is maternally inherited, and so a male clone would quarantine the genetic information he inherited from the domestic egg donor in his body. Biology was here used to delimit when and how the cloning technique could be used. This was not, however, an unmediated biology; it was a biology that had been coconstituted, or coproduced (Jasanoff 2004; Reardon 2001), with the social practices that zoos commonly use, as seen in chapter 1.

Using Ryder's white paper, the San Diego Zoo Global decided that the gaur was not the best animal to clone. Rather, the Society decided to use cryopreserved fibroblast cells from a deceased banteng. This banteng was considered "genetically valuable" because he did not have extensive kin relations within the captive population. However, he died prematurely in the 1970s due to an accident, and had not produced offspring. The level of genetic diversity within the captive population could thus increase by cloning this animal. In addition, the banteng was a male; his offspring would not inherit the domestic mitochondrial DNA inherited from the egg cell donor.

The division of labor established in cloning the gaur was largely followed in the subsequent project to clone the banteng. Cryopreserved somatic cells from the genetically valuable banteng were sent to ACT, where the processes of interspecies nuclear transfer were carried out using domestic cow egg cells. The resulting embryos were sent to Trans Ova Genetics for domestic cows to gestate. However, this time the San Diego Zoo Global sent a team to Trans Ova Genetics that was made up of a veterinarian, a nutritionist, a neo-natologist, a curator, and a researcher.[6] The practices involved in livestock husbandry were deemed insufficient for endangered animals, thus requiring the additional and specialized knowledge of zoo workers in order to care for the newborn banteng. It is impossible to say whether changing the procedures in this way made the difference, but one banteng did survive and is currently on display at the San Diego Zoo. He is the chocolate brown cow in the middle of Illustration 4.1.

The cloned banteng embodies an attempt to develop biotechnologies in a pluralist and more democratic manner. Charis Thompson (2002b) has argued that, in the politically charged environment of conservation, scientific practices must engage multiple perspectives. Ryder explicitly took this approach in writing his white paper. He took the criticisms that people working in the zoo had of the gaur project and addressed

Illustration 4.1: Cloned banteng at the San Diego Zoo
(Photograph taken by Stephanie Miller, reprinted with her permission. Permission also
received from the San Diego Zoo Global to reprint.)

these critiques while rearticulating a different kind of cloning project. Rhetorically, organizing technology development in a pluralist fashion did serve a significant function. Some—but certainly not all—of the most ardent critics of the gaur, African wildcat, and sand cat cloning projects were ambivalent, or even positive, in their remarks about the banteng cloning project.

Embedding Cloning in the Zoo

The cloned gaur, African wildcats, sand cat, and banteng were all created to demonstrate the feasibility of interspecies nuclear transfer. Each animal was also, to some degree, created to generate a spectacle in the popular press, in a manner that—as we saw in chapter 3—has been interlinked with the role of hype in producing funds for not only biotechnology companies but also zoos. For many people working in zoos, these goals have not, however, seemed directly relevant to their current or future work in species preservation. As a result, the cloned banteng was created to also demonstrate how the processes involved in developing a new and experimental technique—along with new and experimental animals—could be articulated within the existing practices of zoological parks. By changing the somatic cell donor, cloning has been articulated with the ex situ preservation practices of the park. The cloned banteng has thus become embedded in the zoological park in a way that Miles, the cloned African wildcat, has not.

I focus here on how the banteng embodies this link between cloning, frozen zoos, and SSPs. A central argument of technology studies has been that engineers and scientists will not be able to create well-functioning technological systems if they do not account for how these systems are embedded in social life (Bijker, Hughes, and Pinch 1987). Technology does not enter society in a linear fashion; the two are instead coproduced (Jasanoff 2004; Reardon 2001). In a similar manner, pluralist technology developers have argued that the social systems enabled by technology need to be made explicit within experimentation. In turn, they have engaged a greater range of actors in the technology development process. I thus focus on the two key sets of actors with which cloning has here been articulated.

Species Survival Plans

Zoos have historically been based on collecting animals from the wild for display. However, since the 1970s zoological parks worldwide have been increasingly focusing on reproducing their captive populations. The website of the International Species Information System proudly proclaims that in zoos "about 82 percent of new zoo mammals are now born in captivity, along with more than 60 percent of birds and a majority of reptile species" (Meyer and Hunt 2008). Here, the zoo animal is not authentic because it came from the wild, but rather because it shares a genetic tie with a species that has not been domesticated. Cloning has been made meaningful for the zoo through pluralist technology development in the context of this set of reproductive practices wherein captive populations of "wild animals" are made rather than found.

This shift in collection practices responds to the colonial history of zoological parks. Collecting and selling wild animals became a business during the nineteenth century through the establishment of colonial networks and the growing desire for exotic animals by zoological societies as well as menageries, circuses, and laboratories.[7] Here, animal catchers would kill adult animals in a pack to capture their young for transportation to other locales. However, World War II seriously disrupted this global animal trade, with Germany (the leading wild animal broker) losing both its colonies and shipping companies (Hanson 2002: 87). As colonized locales gained independence and became nation-states post–World War II, newly founded countries began to restrict the number of animals that could be exported. It became apparent that animal collecting, based in the consumptive, extraction practices of colonialism, contributed to the endangerment of certain species. Meanwhile, the three Endangered Species Acts that U.S. Congress passed between 1966 and 1973, along with the Convention on International Trade in Endangered Species of Wild Fauna and Flora (CITES), increased regulation over wild animal collections and therefore zoos. This included both importing practices and keeping wild animals (Hanson 2002: 166–167).

These disruptions to the wild animal trade coincided with a shift in the way people thought about wildlife in general and zoos in particular. Many European and American zoos became run down and neglected between the 1950s and 1970s,[8] raising public concern regarding neglected animals living inside zoos. This coincided with the rise of environmental movements during the 1960s, some of which were critical of keeping wild animals in captivity. These developments were coupled with the expansion of wildlife programming through television, which changed the way people thought about wild animals.[9] Freedom became a central discourse in framing the meaning of nature in this context, and the lack of freedom became a core criticism of zoological parks.[10] Calls to abandon the zoological park altogether increased as a result.[11]

Since the 1970s, conservation has become the primary mode through which zoological parks have redefined themselves in order to regain public support for the zoo as an institution. The conservation turn amongst zoos is often symbolized by the reorganization of parks from individual animals taxonomically displayed vis-à-vis one another in barred cages to immersion exhibits where animals are displayed in social groups and within settings that are meant to resemble the locales from which they originate. These immersion exhibits are often traced back to Carl Hagenbeck's innovative designs at the Stellingen Zoo in Germany, where moats rather than bars were used to enclose animals to create an image of freedom. In his historical analysis of Carl Hagenbeck, Nigel Rothfels (2002) persuasively counters the assumption that Hagenbeck's design can be aligned with better treatment of animals. Rather, immersion exhibits reflect a change in the idea of nature based in the idealization of freedom.[12] To entertain zoo-goers, Hagenbeck made the containment of animals invisible to create an illusion of freedom that mimicked the aesthetics of wildlife films.

But the display practices of zoos were not the only site to change with the conservation turn. Breeding animals increasingly became a necessity in order to sustain the institution itself (Hanson 2002), rather than one of its many curiosities (Anderson 1998b). At first the focus was on propagating zoo animals, or the "demographic approach" I discussed in chapter 2. In practice this meant that the same animals were often repeatedly bred, however. Some zoo animal populations became severely inbred as a result, which meant that there were many sick and suffering animals within the zoo.

In response, zoos began to manage their animal populations through selective breeding in the 1970s and 1980s.[13] In the United States, Ulysses S. Seal started the first SSP at the Minnesota Zoo for Amur tigers.[14] Here genealogical records were created for the total captive Amur tiger population across North American zoos. Using these records, Seal began to work with other zoos so that the "right" tigers were bred with one another. The right breeding pair was determined by asking which animals are 1) not related to each other, and 2) not well-represented in the captive population. Zoos began to ship their animals to one another so that the right animals were placed in a display with an appropriate mate. People I spoke with referred to this as "genetic management."

Species Survival Plans were increasingly institutionalized across the 1980s and into the 1990s. At the time of writing, the Association of Zoos and Aquariums's website noted that there are three hundred SSP programs.[15] Most zoos worldwide use these management schemes to reproduce their endangered populations, although different countries use different names to distinguish their programs. In this context, zoo animal reproduction is increasingly managed at a regional and even global level. For example, a zoo scientist told me that it is not uncommon for zoo animals to be transported between Australia and the United Kingdom in order to ensure that the proper animals breed. And many people I spoke with told me about a group of German scientists who travel around the world inseminating the right elephants with the right sperm, thus avoiding the transportation of these large animals. So whereas zoos used to rely on a global traffic in wild animal bodies premised upon collection, today zoos rely upon a global traffic in zoo animal bodies and bodily parts that is premised upon selective breeding.[16]

Breeding animals in this way has become one of the primary contributions of zoological parks to species preservation efforts. The zoo creates rather than collects a population of safely held animals in a controllable environment, which can serve as a backup or reserve for more precarious endangered populations that reside in less controllable native habitats. Meanwhile, the zoo can continue to entertain and educate the public, emphasizing the importance of conservation.[17] The scientific identity of the zoo is not exemplified through display any more. Rather, the scientific identity of the zoological park is achieved through the ability to reproduce very small endangered animal populations over time. These knowledge practices are multiply situated. On the one hand, veterinary along with field- and zoo-based animal behavioral research is used to create better spaces, diets, and activities for animals, which can encourage zoo animals to breed of their own accord. On the other hand, genetic and reproductive sciences are used to ensure that "the right" two animals reproduce, using assisted reproductive techniques if needed. Together, these specialisms are developing knowledge that is meant to help with the reintroduction of zoo animals into their native habitats, in an attempt to realize the full conservation potential of the park.

The primary technology that SSPs use to manage captive, endangered animal populations is the kinship chart. This is a record keeping technology, or "memory practice" (Bowker 2005), generally associated with domesticating animals. It tracks the genealogy of one individual in relation to others within the population. SSP organizers use these records to determine which two individuals should breed, and which animals should not be bred. But whereas agriculturalists, pet fanciers, and medical researchers use these records to inbreed animals in order to create genetic homogeneity, SSP managers use them to outbreed zoo animals in order to create genetic diversity. The SSP managers I spoke with described their work to me as follows:

> One thing we can do in zoos is we can manage these animals more intensively. So it's not a randomly breeding population. It's not a matter of who runs into who [in determining] whose genes get passed on and whose don't. We control that. We basically say, "This male needs to breed with this female because their genes are not very well represented in the population."
> *Interview, Species Survival Plan manager and Taxonomic Advisory Group cochair (April 8, 2006)*

> You keep the pedigree, the historical pedigree of all the animals that we know about that have ever existed in the captive population. There's some real slick software that you can use to establish things like the diversity within the population, mean kinship. You use these mean kinship values to help you set up the best pairings, or the pairings that will ideally help you preserve the most amount of gene diversity that you can possibly preserve.
> *Interview, Species Survival Plan manager (October 18, 2005)*

The way these SSP managers describe their work shows that endangered animals living in captivity are already being actively made at the population level. The reproduction of these animals is no longer rooted in the randomness of evolution, but is instead governed through selective breeding. Zoos have coopted studbooks and kinship charts from agriculture and the pet industry here, in order to outbreed animals for the purpose of biodiversity.[18]

However, despite all the selective breeding protocols currently employed, zoo populations will eventually become inbred. Some genetic information is always lost through sexual reproduction, or the combination of two genomes. When animals are confined to a small space within a closed population, each successive generation will lose genetic variation. This process can be delayed. For example, zoos will commonly breed animals later in life so that the time span between generations is extended. If animals are bred at two years of age, the zoo will have created five generations in ten years. If animals are bred at age ten, the zoo will have created two different generations in ten years. Less genetic diversity has been lost within the ten years in the later scenario. Nonetheless, even with such strategies in place, genetic diversity will eventually be reduced to the point where the population is no longer "viable." This is why founders, generally animals collected from the wild, are considered the most "genetically valuable" animals within captive populations. They are not related to any other animal within the zoo, and can only create diversity through their reproduction. Founders are thus necessary for zoo populations to persist into the future.

Zoos need to have new genetic information brought into their populations. This has traditionally occurred by collecting wild animals from their in situ habitats, something that people I spoke with unanimously believed should be avoided. Keeping charismatic endangered species in habitats is crucial to the politics of conserving diverse ecosystems. Because people want to preserve endangered species, they will support legislation that protects the habitats in which these animals live. From a conservation perspective, this allows many types of less charismatic animals and plants to also flourish. A full range of different species are given the space to coevolve into the future. As a consequence, taking charismatic wild animals out of their habitats to display in a zoo draws into question the conservation value of zoological parks. Cryopreservation offers a possible solution to this problem.

Frozen Zoos

While SSPs have sought to reproduce rather than collect captive populations of endangered animals, "frozen zoos" have been developed to collect and keep endangered animals in a different way, as cryopreserved cells rather than fully formed animals. The Frozen Zoo™ in San Diego is probably one of the longest standing and largest collections of cryopreserved cells from wild and endangered animals. Kurt Benirschke started this collection in 1965 in New Hampshire; it was institutionalized at the San Diego Zoo in the 1970s. Benirschke and other scientists initially developed this collection by opportunistically preserving fibroblast cells from ear notches that were taken from hoof animals, per legal requirements.[19] These collection practices quickly expanded to include cell samples taken from zoo animals after death, across a range of different species. The majority of cells continue to come from zoo animals today. However, some field conservationists—who routinely capture wild animals to take biological samples, administer a tracking device, and then release the animal—are also taking sperm samples from males for donation to frozen zoos.[20]

The Frozen Zoo™ was originally created to do genetic analyses of wild and endangered animals and populations. Benirschke described it to me in this way:

> It's mostly used as a repository, for the people who are interested in getting DNA from species. That's what is being requested of samples most of the time. It's a repository for the future. That's basically, exactly, what it is. We're trying desperately to get very fresh material from chimpanzees that are dying.
>
> *Interview (September 15, 2005)*

Here, Benirschke describes the value of frozen zoos as repositories of genetic information that serve not only scientists today, but also scientists of the future. Endangered animal cells are cryopreserved in order to keep a genetic record of the species living on the planet today, which may not persist into the future in embodied form. The goal is to remember the genomes of species before they go extinct. The idea is that the frozen zoo can serve a future generation of scientists who will lack the requisite research materials for asking questions about biodiversity. The cryopreserved cell of the extinct bird embodies this idea.

When reproductive scientist Barbara Durrant joined CRES in the 1980s, she did not think the frozen zoo should simply be acted upon as an archive; she thought that cryopreservation also had a reproductive function in the preservation of endangered species (Interview, May 16, 2006). Cryopreserved sperm could, for example, be quite useful to the SSPs that were developing at that time. As such, Durrant expanded the Frozen Zoo™ to include not only fibroblast cells but also sperm and, later, embryos. If zoos want to bring endangered species "back from the brink," the frozen zoos could contribute to this effort by bringing individual animals back from the bank.

By linking frozen zoos with SSPs, endangered species are increasingly being thought of as including both embodied individuals (where genetic information is materializing) and cryopreserved cells (where genetic information is suspended in a liminal state). For example, Durrant told me that the new Zoo Information Management System (ZIMS) will incorporate semen collection, semen evaluation, embryo collection, and embryo classification in its database of the world's zoo animals. Sperm and embryos are thus considered protoanimals in the kinship charts that SSP managers use. Cloning potentially adds another kind of cell that can be included as an "individual." Where fibroblast cells have been thought of as scientifically productive, in that future scientists could use these cells to ask genetic questions about lost species, post-Dolly the Sheep these cells are increasingly being considered reproductive.

In the discussions I had with people while conducting this research, it appeared that one of the greatest impacts interspecies nuclear transfer has had on zoos is in legitimating and expanding cryopreservation. For example, the possibility that life could emerge from collections of fibroblast cells has already changed the practices of some zoological parks. Philip Damiani (Interview, July 1, 2005) told me that whereas many zoos previously cryopreserved tissue so that DNA analysis could be conducted, many are now also preserving cell lines that could potentially be used in conjunction with techniques ranging from artificial insemination to interspecies nuclear transfer.

Re-Envisioning the Spectacle

As discussed in chapter 2, there is very little spectacle surrounding the cloned banteng in the San Diego Zoo. In the park, this animal looks like a rather ordinary cow. Both times I visited this animal, the clone and his two female companions were all resting in the shade. There was little interest in these seemingly contented animals by fellow zoo-goers, who generally seemed more intrigued by the polar bear whose enclosure was across the road.[21] As such, cloning does not contribute to the visual spectacle that is central to the education and entertainment mission of zoological parks. Nor is the frozen zoo very spectacular to see. Indeed, it requires a number of symbolic gestures to render its work meaningful to visitors, such as the pictures of animals attached to its locked door.

From the zoo's perspective, the cloned banteng was not intended to be spectacular to zoo-goers. Nor was this animal intended to be spectacular in the popular press, at least as far as San Diego Zoo Global was concerned. Indeed, Oliver Ryder told me about an interaction he had with a reporter shortly after the banteng was born. He recalled telling the journalist why the banteng was chosen for the cloning experiment, and how this animal would contribute to the zoo's efforts in preserving this endangered species. At this point, the reporter interrupted him, asking if the banteng was extinct. When Ryder said no, the reporter apologized and said he would need to call his editor; his editor thought that an extinct animal had been cloned, and might not want to run a story about "just" a cloned endangered animal. After a little laugh, Ryder told me that he was happy to have had that encounter; it showed him how normalized their work had become. Whereas the gaur, African wildcats, and sand cat were meant to be spectacles in the mass media, Ryder was quite happy to find out that the banteng was not spectacular enough to make the news.

The spectacle of the cloned banteng was instead directed toward SSP managers. After the birth of the banteng, Oliver Ryder and Philip Damiani were asked to attend the Taxonomic Advisory Group meetings for hoofstock in North American zoos. Here they described how interspecies nuclear transfer works and how it could be applicable to the zoo. According to people I spoke with who attended that conference, the presentation focused on the ways in which cloning could be used to regenerate the cells of genetically valuable individuals that are preserved in frozen zoos.[22] This would allow SSPs an additional set of options when managing their populations. According to Ryder, this also allowed him to learn about how the Frozen Zoo™ could forge a closer relationship with the Species Survival Plans. After this presentation, the Taxonomic Advisory Group decided to continue this cloning experiment, and have since included this animal in the banteng SSP.[23]

The cloned banteng, along with a vial of semen from a now deceased banteng, were considered the two most valuable "individuals" in the captive banteng population while I was conducting this research. This designation was based on the extent of genetic diversity that will exist within the captive population if these animals successfully reproduce.[24] The banteng population manager for the North American captive population told me that, if the banteng reproduces, the SSP will reach its goals for genetic diversity in this captive population more rapidly than previously thought possible. But the Taxonomic Advisory Group also recognizes that this is very much an experiment, and there may be problems with the cloned banteng and/or his offspring as a consequence of his mode of reproduction.[25]

Pluralist Technology Development

Despite sustained uncertainties, the banteng has come to embody one of the most persuasive use values of cloning for ex situ species preservation to date. Many critics of technology development have been supportive of the banteng cloning project. What has made this particular articulation of cloning relatively sturdy?

I noted in chapter 2 that Miles, the cloned African wildcat, lives to demonstrate the fact of interspecies nuclear transfer. He is an object in the Latourian (2004b) sense, in that his birth unified the gathering of humans and nonhumans that took place within the cloning process through which he was born. But as a consequence, the life Miles was born to live has remained uncertain. We could say that Miles ceased to be a "matter of concern" (Latour 2004b) when he was born. Bruno Latour developed the idea of a matter of concern as part of his argument for a politics of things. Here Latour has emphasized that "things" are not only entities that are out there in the world but also assemblies of humans and nonhumans that gather in dealing with and settling disputes. Latour notes that what makes things sturdy is that "its participants, ingredients, nonhumans as well as humans, not be limited in advance" (Latour 2004b: 246). Drawing on Latour's work, I would say that pluralist technology development enrolls a greater number of actors in the gathering that is required to make and sustain an endangered animal. And this gathering does not end once an animal is born, but is instead maintained across the animal's life. It is for this reason that pluralist technology development has experienced greater stabilization. The banteng has gathered a more diverse set of actors in dealing with and settling disputes about cloning in zoological parks.

Importantly, this repudiates the argument that cloning was first developed and proven with the gaur, African wildcats, and sand cat and then applied to the banteng.[26] Quite simply, the cloned banteng was born around the same time as the African wildcats and well before the sand cat experimentation started. Scientists at ACRES could have organized their experimentation in this way, by cloning a "genetically valuable" sand cat, had they wanted to. Rather, scientists at ACRES actively decided not to because this would have slowed down their scientific research. But more importantly, the banteng embodies a different way of doing science. Cloning was here explicitly articulated with other techniques and practices within the zoo in order to experiment in a more pluralist way. The experiment was organized as a different kind of "matter of concern" (Latour 2004b). The goal was never to establish interspecies nuclear transfer as a fact, in order to make animals for the zoo. Rather the goal was to create and sustain a gathering of humans and nonhumans in and through the processes of cloning itself in order to remake zoo populations.

The social theorizing from STS that I draw upon here is continuous with changing approaches to technology within some parts of environmentalism. As previously discussed, environmentalists have traditionally been skeptical of technological solutions. However, some are beginning to argue that technologies must be embraced in order to reckon with the magnitude of environmental problems today, in the context of global warming and the suffering this is causing to both human and nonhuman populations. Stewart Brand's *Whole Earth Discipline* (2009) articulates most clearly this pragmatic approach to technologies. However, Brand does not argue that environmentalists should simply accept industry's articulation of any given technology in a linear fashion. He instead argues that environmentalists should become part of the technology development process early on. Brand (2009: 204) states:

> [T]he best way for doubters to control a questionable new technology is to embrace it, lest it remain wholly in the hands of enthusiasts who think there is nothing questionable about it. I would love to see what a cadre of dedicated environmental scientists could do with genetic engineering. Besides assuring the kind of transparency needed for intelligent regulation, they could direct a powerful new tool at some of the most vexing problems in our field. . . . The true nature of any new technology can be learned best from what enthusiasts do with it. Critiques based on the experience of practitioners, rather than on ideology or theory, have real bite.

It is well established that technologies have politics (Winner 1980), without being deterministic (Wyatt 2008). In this context, many corners of both science studies and environmentalism are arguing that critiques of technologies are not enough. Rather, technologies should be engaged in order to reshape their politics.[27] Cloning the banteng represents a site where technological doubters embraced a contested technique, taking it into their own hands—and thus out of the sole control of enthusiasts— in order to refashion its politics. The banteng does not embody the idea of a linear relationship between science and society. Rather, the banteng explicitly embodies the politics of coproducing technologies and social orders in tandem, so as to make nature differently.

Cloning as a Space-Saving Technology

The banteng cloning project was not articulated with the problem of species boundaries at the forefront. Nor was it primarily motivated by a desire to reorganize reproduction across species lines. Rather, this artic- ulation of cloning was centrally organized according to the problem of space. Zoos are necessarily small spaces which can only house a small number of animals. Meanwhile, the planet is also increasingly being understood as a much smaller place (McKibben [1989] 2003: 1), with limited scope for other species to evolve. In this context, zoos cannot collect animals from the wild; they must reproduce their own popula- tions. The banteng embodies a solution to these twin problems of space.

The ways in which cloning can serve as a space-saving technology are rooted in the ways this technique retemporalizes the body. In her historical study of cell culture, Hannah Landecker (2007) has argued that developments in biotechnology across the twentieth century have been premised upon realizing and exploring the plasticity of biological time in living organisms. Cloning is part of a century-long experiment in starting, stopping, and restarting biological time (Landecker 2007: 227). If somatic cell nuclear transfer has done anything, it has chal- lenged the notion that cells make irreversible changes during develop- ment from which they cannot turn back; cellular time is not linear but can instead be reversed (Franklin 1997a, 2007b).

In the case of the banteng, the plasticity of biological time has been operationalized as part of a biopolitical formation that links cryopreserved cell lines, embodied individuals, and populations into a management scheme. Death has historically been the limiting force to biopower (Foucault 1978). Indeed, Kathrin Braun (2007) has argued that linear biological time has been the basis for biopolitical regimes across the twentieth century. Cloning and cryopreservation are enacted to bypass the constraints of aging, decline, and death. The Species Survival Plan manager for the banteng commented during an interview:

> You can bring dead animals into the [kinship] analysis. So you could go and incorporate all the animals for which there are frozen gametes and bring them into your living population. So say if we did an AI [artificial insemination] using sperm from this animal that is currently dead, what would that do to our population? And you could do the same thing if you wanted to bring these other animals for which there are tissue samples.
> *Interview, Species Survival Plan manager (October 18, 2005)*

The cloned banteng shows how the practice of delaying pregnancy in order to preserve genetic diversity could be extended with somatic cell nuclear transfer. Reproduction can occur postmortem. An animal can reproduce when it suits the needs of the population, unfettered by the aging process that shapes the possibilities of individuals. Cloning allows for biological time to be folded back and thus potentialized for the future. Death is no longer a limit to a biopolitical regime rooted in preserving the diverse life of nature.

But what motivates this reworking of biological time in the zoo is centrally the problem of space as opposed to the problem of time.[28] In Ryder and Benirschke's (1997) initial imaginings regarding the potentiality of somatic cell nuclear transfer, they started with the fact that cryopreserved cells take up very little space when compared to a living animal. Species preservationists today know that a minimal number of animals are required to sustain the genetic diversity of in situ and ex situ populations, which in turn requires a minimal amount of land. But if living animals and suspended genomic information were understood as part of a single population, fewer embodied individuals would hypothetically be required. And this means that less space would be required to sustain a viable population. For Ryder and Benirschke, relocating genetic diversity in this way would free up already limited herd spaces, allowing for the preservation of a greater number of endangered species.[29] Pluralist technology development has thus been rooted in an imaginary that seeks to work around the spaces in which animals live in order to reproduce endangered populations in new ways.

Working around space requires a genetic definition of endangered species. Zoo animals are wild not because they came from the wild, but rather because they have a genealogical relationship with wild animals. The male is an endangered animal because his species relation with a wild animal can be sustained into the future, while his species relation with a domestic animal will end with his death. Genetics defines the species in this instance. This is why both pluralist technology developers and zoos alike have been generally concerned about changing the genome of endangered animals. Unlike technology developers, they do not want to remake endangered animals from the inside out. Rather, they want to keep the genome of these animals the same, as this is what links captive animals to the wild.

As an institution, the zoo is particularly concerned with patrolling the boundaries between nature and culture. Like other groups organized around environmental issues, zoos rely upon the modern belief that wild species biologically preexist human intervention; in this social milieu "species" is not simply a taxonomic convenience but a biological fact used to direct and delimit social practice. In turn, the genetic basis of species is crucial for zoological parks. To conduct life science research with wild animals and endangered species, these entities need to be able to stand as genetically pure. To educate the public on the importance of conservation, children and their parents need to see "real" endangered animals, and genetic lineage provides a means to assert the authenticity of wild animals bred in the zoo. Finally, the focus on the genetic basis of species diffuses the idea that wild animals are kept in captivity for human pleasure, thereby legitimating the zoo as a site of good husbandry that can recapitulate species as they exist in the wild. Purifying the genetic basis of endangered species is highly significant in this particular situation, and is in turn heavily patrolled.

Transposing the bodies of different species, and therein mixing the genomes of different kinds of animals, is experienced as an affront to the values of the zoological park. Latour (1993) has argued that the more concerned social groups are with maintaining purity, the more attention they will pay to hybrid forms so as to control and delimit such entities. By emphasizing the hybridity of DNA embodied by cloned endangered animals, zoo scientists were able to delimit and contain such impurities. Pluralist technology development provided a way to remake endangered populations that are "natural" according to a genetic definition of species.

That said, genetics is not the only way of defining the nature of species. In delineating memory practices as a kind of scientific work, Bowker (2005: 7) has emphasized that the traces we leave are not the same as what we are. Rather, the traces that make up an archive represent a negotiation between the archivist and future auditors, whose work is being both envisioned as well as acted upon. The frozen zoo represents a way to preserve fragments of endangered species as they exist today, so that future scientists will have the required research materials to continue to study and know species that we fear may go extinct. It also represents a way to genetically manage endangered populations, to help avoid such a loss. This is why genetic traces of endangered species like chimpanzees are being preserved, so that these species can be both known and saved. But the genetic traces of animals like chimpanzees are not the same as chimpanzees. Indeed, part of the reason why chimpanzees intrigue people is because their behaviors differ so much across different spaces (Rowell 1972). The future of zoo science is thus refracted through a molecular lens within the frozen zoo, and this lens leaves physiological, behavioral, and environmental questions out of view. The constitutive role of space is thus actively ignored so as to purify not only cloning, but also the zoo. The next chapter will explore in more depth these "genetic values," which are so central to zoological parks today.

5

Genetic Values

The placard outside the banteng's enclosure at the San Diego Zoo states that one of the animals on display is a clone, as noted in chapter 2. It begins in a rather typical manner, asserting that this endangered bovine is a "true species." This differentiates these animals from other, presumably domestic "breeds" of cow. The placard continues to describe the phenotype of the banteng, largely through the lens of sex differences. Here the zoo visitor learns that the dark, chocolate brown individual with horns is a male, while the lighter brown individuals without horns are females. These are common lessons at the zoological park. Zoo visitors frequently learn to appreciate "wild" animals by seeing species that look different from other, more common kinds. These animals are then often discussed in terms of differences that exist within the species itself. Both are lessons in the importance of sex, wherein sexual reproduction and sexual dimorphism sustain species difference.

Rather blandly, the placard continues by stating that the male in the enclosure is part of an evaluation project to see if animal cloning could be a tool in species preservation. The zoo visitor learns that the male in the display was created from a "genetically valuable" banteng that lived in the zoo three decades earlier, during the 1970s. The lesson in sexual dimorphism gains an additional use value here. It allows visitors to see which banteng in the group is the clone, having inherited his genetic value from a single, long-deceased ancestor. However, the placard ends before addressing what seems to be the obvious next question. How does cloning—as a mode of asexual reproduction—assist a species preservation project that is rooted in the production of difference through sex?

This chapter examines how cloning reproduces and sustains the centrality of sex in the biopolitics of the zoo.[1] The cloned banteng was made and born in order to be genetically valuable, which requires that he not only regenerate lost genetic information but also reproduce that genetic information. The banteng's genome must "mix" (Franklin 2007b) with the captive population. Importantly, the zoo is not envisioned here as a place populated by hundreds of clones, carbon copies of endangered animals living in the wild. Contra dominant discourses on cloning, wild animals are not being mass produced in any zoological park. Asexual reproduction has not replaced sexual reproduction here. Rather, the zoo is envisioned and enacted as a place that uses available technologies in order to manage sexual reproduction, and thus sustain a deftly balanced genetic composition within a limited amount of space. The goal is to shepherd endangered populations, to put them on the right genetic track.

The cloned banteng thus embodies not only a general scientific ethos based in pluralist technology development, but also a set of values that prioritizes genetic definitions of species. Unlike the fascination with transforming the biologies of endangered animals, these genetic values circulate amongst zoos more generally and are accepted. To understand how and why zoos have come to value genetics in this way, I put this idiom next to Harriet Ritvo's (1995) concept of "genetic capital," through which she has conceptualized early breeding practices in agriculture. While there are important differences between the "domesticatory practices" (Cassidy and Mullin 2007) of agriculture and species preservation, both work by minimizing space and place in delineating the nature of animals in favor of genetic definitions. I conclude by considering how nature and culture are connected in these genetic values, and the kinds of relations with endangered species that are therein enabled.

From Genetic Capital to Genetic Value

Selectively breeding plants and animals is an area where there has been sustained interest in producing capital-intensive bodies by exploiting biogenetic relations. British agriculturalist Robert Bakewell (1725–1795) is most famously known for his role in elaborating this system of valuing animals in agriculture. Agriculturalists of Bakewell's time generally located the value of animals in the health and size of herds, wherein individual animals were considered more or less exchangeable.[2]

Bakewell instead looked to the reproductive potential of specific animals that displayed highly desired traits as the source of value.[3] Ritvo (1995) has described and characterized these breeding practices, arguing that Bakewell created "genetic capital" by transferring the value of an animal from the location and environment in which it dwelled to the biogenetic genealogy from which it came.[4]

To understand the significance of genetic capital, it needs to be emphasized that the idea of heredity was only just developing during Bakewell's lifetime. The then predominate thought was that individuals—human and animal alike—develop their characteristics and traits through "generation."[5] Staffan Müller-Wille and Hans-Jörg Rheinberger (2007: 3–4) have noted that nature and nurture as well as heredity and environment were not viewed as oppositions in this context. Characteristics were instead linked to the physical and emotional spaces in which humans and animals came into being; the environment was thus assumed to constitute the human and nonhuman beings residing within it. Shared space explained resemblances between different individuals.[6]

Bakewell's creation of genetic capital through selective breeding was part of what Müller-Wille and Rheinberger (2007) have called the "epistemic space" of heredity, which was emerging during the eighteenth century. In their cultural history of the practices informing this emergence, Müller-Wille and Rheinberger have argued that heredity initially developed across varying knowledge regimes independently (which included medicine, natural history, and breeding), and in a rather fragmentary fashion.[7] These developments built upon legal and political concerns with inheritance, which was in part spurred on by the creation of mobile as opposed to landed property.[8] The mobility of new forms of wealth in turn coincided with—and possibly gave rise to—the mobility of plants and animals with botanical gardens, menageries, and the trading practices of breeders. The ability to move plants and animals to new spaces, while retaining their physical features, raised new questions that helped form "heredity" as a concept. In this context, Müller-Wille and Rheinberger (2007: 19) have noted that natural history, premised upon description, was among the first sites to consider heredity from an experimental point of view. Biology came to be understood as separate and independent from the environment.

In partaking in the formation of "heredity" as an epistemic space, Bakewell also capitalized on the reproductive potential of prized male sheep whose traits could now be spread through sexual reproduction. Ritvo (1995) has described how Bakewell would loan out individual, male animals who embodied highly desired traits for a breeding season, in exchange for a stud fee. Purchasers bought the reproductive potential of these animals, and their transformative powers for herds. Ritvo (1995: 417) has described Bakewell's logic of capitalization as follows:

> Bakewell claimed that when he sold one of his carefully bred animals, or, as in the case of stud fees, when he sold the procreative powers of one of these animals, he was selling something much more specific, more predictable, and more efficacious than mere reproduction. In effect, he was selling a template for the continued production of animals of a special type; that is, the distinction of his rams consisted not only in their constellation of personal virtues, but in their ability to pass this constellation down their family line. . . . Thus it was possible for a disciple like George Culley of Durham to transform his own flocks by hiring a ram from Bakewell each year.

As such, Bakewell created a system of value that was based on the biological as opposed to environmental inheritance of traits (Ritvo 1995: 415–416). In turn, agriculturalists bought not simply an animal but also the reproductive potential of that animal. As a consequence, people could have a flock of animals that looked quite different well within the agriculturalist's lifetime, something that would have been impossible according to the logic of generation.

By selecting particular rams, Bakewell was commodifying not only a prized individual but also the genealogical potential of that individual, and thus the continuance of that lineage into the future (Ritvo 1995: 415). Record-keeping practices, specifically studbooks, had to be produced in order to demonstrate this valuation system. Ritvo (1995: 419) has pointed out that these records were hard to come by, and were quite possibly inconceivable in the eighteenth century, when individuals within herds were understood as interchangeable parts. As such, Ritvo has argued that the creation of studbooks had to follow the practice of selective breeding. However, this practice was also simultaneously constitutive of breeds. Studbooks thus helped to both organize agricultural production through selective breeding and prove that genetic capital had in fact been created.

Genetic Value

The selective breeding practices developed in agriculture have provided the basis for creating genetic value in zoos, where techniques like cloning and cryopreservation are also used in relationship to kinship charts and studbooks. That said, the discourse of genetic capital has also been revised as these practices have settled down in zoological parks. The shorthand of "genetic value," which people in zoos commonly used in our conversations, embodies both the connections and fissures between selective breeding across these institutions. To understand the salience of genetic values within zoos, I compare and contrast "genetic capital" to "genetic value." In other words, I ask how selective breeding has been reworked as it has been transposed from agriculture to the zoo.

Like genetic capital, creating genetic value in zoos has relied upon relocating the value of an animal from the space in which the animal lives to the genealogical relations from which it came. Historically zoo animals were valued because they came from wild spaces. Hunting, collecting, and transporting were the primary practices underlying this valuation system. The incorporation of selective breeding protocols into zoos is one site where this valuation system has been problematized. Zoo animals are no longer treasured because they come from a faraway place. Indeed, the whole logic of genetic value has relied upon the idea that endangered animals have *not* been removed from their habitat in

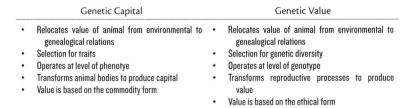

Genetic Capital	Genetic Value
• Relocates value of animal from environmental to genealogical relations	• Relocates value of animal from environmental to genealogical relations
• Selection for traits	• Selection for genetic diversity
• Operates at level of phenotye	• Operates at level of genotype
• Transforms animal bodies to produce capital	• Transforms reproductive processes to produce value
• Value is based on the commodity form	• Value is based on the ethical form

Figure 5.1: From genetic capital to genetic value

order to be placed in the zoo. Rather, the zoo animal is valuable because of its genealogy in a wild species that is considered endangered.

However, selective breeding has traditionally focused on reproducing desired physical traits embodied by certain individuals. This has meant that males and females from the same line are inbred in order to increase the chance of producing offspring embodying the same, desired characteristic. Having a particular stature or type of coat, being able to run fast or produce a lot of milk are all traits for which agriculturalists and pet fanciers have selected animals. Selective breeding has thus been a way to both reproduce and amplify desired traits among domesticated animals. And it is according to this logic that cloning has been understood as the ultimate form of inbreeding. Because genes thought to give rise to traits are recapitulated and regenerated, rather than mixed, cloning has been understood as a more efficient means to inbreed animals.[9] An industry scientist I spoke with summarized this as follows: "So for instance if they had a cow that might be, say for instance a Holstein cow that produces a lot of milk, the idea was to clone that animal so that the producer would potentially have a small, elite herd of identical animals" (Interview, July 1, 2005). Here we see the idea that cloning creates a copy of not only the genome, but also the phenotype of a valued animal. As such, cloning reproduces and extends the logic of genetic capital originally developed by Bakewell. Producing identical cows that all embody the same, highly desired phenotypic trait represents an intensification of genetic capital.[10]

In sharp contrast, SSPs are not using selective breeding techniques to reproduce animals with particular phenotypic traits. Rather, zoos want to reproduce individual genomes that are not represented in the captive population. So whereas the kinship chart is used to inbreed agricultural animals to produce physical traits, SSPs use kinship charts to outbreed zoo animals in order to reproduce genetically diverse, captive populations. Thus, while a purebred is the embodiment of genetic capital in agriculture and (some) companion animal industries, purebreds embody a core problem in zoos, one that selective breeding is meant to avoid. A speaker at the 2006 Taxonomic Advisory Group meetings for endangered cats made this difference clear when she adamantly remarked: "I often have to remind people that purebreds are all those sick, deformed, and unproductive animals we want to get *out* of our zoo populations." As such, the goal is not to improve how a population of animals looks or acts by intensifying certain traits. Rather, zoos want to ensure that their animal populations live healthy lives into the future, which requires populations to be genetically diverse.

Cloning thus does not garner value in the zoo because phenotypic traits are copied; rather, genotypic information is collected and distributed in new ways with interspecies nuclear transfer. Genetic information lost through death can be found again. And this has facilitated new imaginaries in which wild animals, or founders, could be collected as cells rather than fully formed animals. An industry scientist articulated this imaginary as follows:

> It's very difficult to import animals into the United States. It's also very stressful and a lot of animals die during transport. I think it would be great to bring in new genetics just via cell lines. And leave the animals back in the wild where they belong, instead of having to capture an animal or two to bring it back to the zoo. This serves two purposes. It would leave the animals back in the wild where they belong. It would also allow for further habitat conservation because we need to preserve the habitat for those animals to be there. Using cell lines we could potentially increase genetic diversity in our captive population, by bringing in so-called new individuals without removing the original animal.
>
> *Interview (July 1, 2005)*

Here we see the hope that the difficult and rather dangerous practices involved in transporting wild animals could be obviated by shipping cells as opposed to animals. Wild animals would thus be kept in their in situ habitats, "where they belong." But the zoo would still be able to sustain itself by getting new genetic information into its populations. Importantly, this imaginary is grounded in contemporary discussions amongst zoos and field conservationists. At the Taxonomic Advisory Group meetings for endangered cats in 2006, I saw a reproductive scientist and field conservationist discuss the possibility of a similar kind of collaboration. Here the field worker would collect sperm from the cats he was monitoring in situ. These cells would serve as founders for the zoo. Cloning and other assisted reproductive technologies allow wild animals to be collected by the zoo again, but ethically, as cells rather than fully formed animals.

While genetic capital is rooted in improving individual animals—and therein a population—for the purpose of capitalization, genetic value is about preserving a captive population for the purpose of conservation. In this context, the mechanisms underlying genetic value are not focused on producing capital in zoos, as seen in agriculture. Zoos do not buy and sell animals for profit. Nor do zoos charge stud fees to breed animals for a season, as Bakewell did. Rather, there is something far more like a "gift economy" (Titmuss [1971] 1997) operating in zoos, one that is overseen and enforced by the SSPs and the Association of Zoos and Aquariums. A zoo scientist I spoke with summarized this gift economy as follows:

> In the conservation world we tend to manage these populations. For example, you may have fifty different institutions in the U.S. that all have white rhinos but they're managed as one population. And they're all moved around to different zoos to maximize breeding. So we don't tend to think of these animals so much as "our" animals. And that's because it doesn't usually translate into a commercial profit.
> *Interview (April 17, 2006)*

Selective breeding has not been used with the primary goal of generating capital in zoological parks. And this is because breeding white rhinos does not normally translate into profit within the park.[11]

If the goal is not to generate profit but rather value in the zoological park, what kinds of value is "genetic value" meant to produce? Value here refers not to the monetary worth of an item in exchange, but rather to the mutually cared about beliefs and practices that occur within a social space.[12] Zoos have exchanged their old collection practices in favor of selective breeding protocols in order to become ethical and valued institutions. They have taken up the techniques associated with domestication in response to public concerns about zoological parks. Zoo animal populations have had to become "self-sustaining" (Benirschke 1986) so that they can continue to educate publics without reducing in situ animal populations. So, whereas genetic capital has been pursued with the goal of capitalization at the fore, genetic value has alternatively been elaborated as a basis for ethical practice. Genetic definitions of species have enabled the zoo to transform itself into an ethical institution that assists rather than hampers conservation efforts. This is why genetic definitions of species are in turn deeply valued within zoological parks. It is no wonder that many zoo workers I spoke with were worried about, and wanted to contain, the hybrid genetics of animals produced through interspecies nuclear transfer. Genetic, as opposed to environmental, definitions of endangered species have to be privileged for the zoo to become the kind of ethical institution that it aspires to be. This is a strategic form of genetic essentialism.[13]

The Political Economies of Genetic Values

Genetic value may not be organized with intensive capitalization as its primary goal. That said, one of the more common statistics I heard from people in zoo worlds is that every year, more than 100,000,000 people visit zoos, aquariums, oceanariums, and wildlife parks in the United States alone.[14] This outnumbers attendance at all football, baseball, and hockey games combined. In this context, wild animals are understood as commodities. And so while breeding zoo animals may not gener- ate capital directly, the practice does allow the zoo to reproduce itself at an institutional level and thus generate capital. As Kaushik Sunder Rajan (2006: 41) has noted, the word value is "one of those nice double- jointed words that always already imply two different things. On the one hand, 'value' implies the market value that gets realized through the processes of exchange. On the other, it implies the nonmarket val- ues that might be called . . . ethics."[15] With this in mind, it is important to emphasize that genetic value pushes on the ethical side of the word, but nonetheless does have financial implications for the park. It is thus important to consider the political economies of genetic value on their own terms, as distinct from their agricultural contexts.

The political economies of genetic value are interlinked with the ways in which animal bodies and bodily parts are exchanged. These exchanges may be a site where the wealth of some is reproduced at the expense of others. Catherine Waldby (2000, 2002; Waldby and Mitchell 2006) coined the term "biovalue" to refer to the ways in which bod- ies, bodily fragments, and biological processes are reformulated to produce surplus "yields of vitality" in bioscience and biomedicine. She has argued that biological fragments often come from marginalized people's bodies. On this basis, Waldby posits that biovalue is a key site where long-standing social hierarchies are reproduced.

Waldby's characterization represents a partial truth in the matters of exchange where endangered animals are concerned. Zoological parks without the resources required to fund expensive biomedical facilities are asked to donate tissue samples to zoos with more resources. Ownership of cells, and any potential animals that might result from those cells, is thus transferred from one zoo to another in the process. Meanwhile, nations with less capital are requested to provide tissue samples to zoological parks located in more wealthy nations. Similarly, ownership of those cells, and any potential animals that might result, are transferred unless stated otherwise.

Both zoos and nations have intervened in these power relations by restricting the mobility of cells and delimiting their use. For example, some zoos are stating that cells donated to a frozen zoo at another institution cannot be used in order to create an animal, thus maintaining their ownership rights.[16] Meanwhile, some nations are refusing to exchange cells in order to maintain ownership of their animals. For example, a member of the Taxonomic Advisory Group for endangered cats told me about his experience in trying to use assisted reproductive technologies in order to establish a captive population of Brazilian ocelots (a small endangered cat) in North American zoos. As part of this conservation program, U.S. zoos contributed financial resources to preserve Brazilian ocelots both in and ex situ. In exchange, U.S. zoos wanted to establish a Brazilian ocelot population in North America using frozen embryos that were cryopreserved in Brazil. According to this population manager, Brazil would have remained the official owner of any animal created from these embryos. But by establishing a captive population in the United States, the amount of "space" available for this endangered species would have expanded. However, the material exchanges that shaped this collaboration came into question when a new government came into power in Brazil. According to the population manager, this government was concerned that such exchanges constituted a kind of biopiracy and refused to ship the embryos to the United States. He commented during the interview:

The new government came in and said: "We don't agree. This isn't something we want to be involved in."... It's this idea that Americans want to take our genetic resources to the United States. And you can tell them about all these other things, but all they see is the United States taking things out.

Interview (April 8, 2006)

These examples demonstrate how genetic value has at times been discursively interlinked with biopiracy, a discourse that has largely developed in the context of anticolonializing social movements and in academic research. Here activists and academics alike have articulated concern over the ways in which American and European biomedical and biotechnological organizations have extracted genetic resources from countries that are materially rich but financially poor. For example, plants have been extracted as part of pharmaceutical production while human bodily parts (e.g., cells and blood) have been taken for the purpose of biomedical knowledge production.[17] Local knowledge regarding those materials is often extracted in the process as well. However, pharmaceutical and biotechnology companies have been able to erase local claims by privileging science in intellectual property law. Laura Foster (2010, 2011, 2012) has pointed out that this is a key site where the authority of "Western" science and law together reproduce unequal relations that are rooted in a colonial history.

The modern zoological park has been rooted in the history of extracting material resources from colonized spaces. As such, this institution has been easily interlinked with the processes of biopiracy, and correspondingly resisted on this basis. For example, the field conservationist I interviewed had this to say about a particular zoo in the United States:

That zoo has a reputation of acquiring exotic and rare species from all over the world to exhibit, from which they make a lot of money. That becomes a danger, when the reproductive technologies become allied with the exhibition [practices] of the zoo. [People in] third world countries—I know some of them personally—have taken the view "why should we give our rare and endangered animals to [Name of Zoo]. Because it only—it's for their profit. Unless we see some way in which we too benefit, then no."

Interview (April 25, 2006)

Genetic value seeks to remediate the colonial history of the zoo. Rather than extract animals from their habitats to be put on display, zoos instead remake their own populations. In doing so, the zoo remediates its colonial history vis-à-vis wild animals. Genetic value does not, however, remediate the global inequalities between different people and nations that have resulted from colonization. The profitability of zoological parks in North America and Europe relies upon exchanges in animal bodies and animal body parts that reproduce economic inequities. An articulation of cloning that explicitly addresses systemic inequalities between people is yet to be organized, but it would likely be an extremely interesting project.

That said, the question of power and politics also exceeds that of ownership and corresponding concerns about who does and does not benefit financially from such exchanges. Questions about who has the right to care for endangered species are also at stake. This necessarily entails questions about how people want to interact and live with other animals. For example, the field conservationist quoted above stated that he was not only worried about questions of ownership, but also with questions about who has the right to decide how people are going to live with other kinds of animals into the future:

> So if the technologists had the technology to be able to stuff our freezers with the full range of genetic diversity we've got all over the world and they can pull it out of the freezer and re-create it at any one point, what does it overlook? It overlooks the fact that they have made themselves the god of the new creation. There are a large number of countries, and particularly Third World countries, which have conserved their wildlife whereas North America hasn't; it's lost most of its big and dramatic animals. We're very dedicated to conserving them in the natural setting and to living with them as a part of our heritage. We do. We live alongside the species and we take a very dim view of people who would say, "Well, don't worry about your species, you might fail. We can pop them in the freezer and we'll pull them out and give them back to you." We don't want the technologist to be the arbitrator of our heritage and future any more than they were in the past. So I think there's an issue of control of your own—just as you would like to see the control of your

own reproductive future, so in many countries we would like to see the control over our own biodiversity future from the past through the present to the future.

Interview (April 25, 2006)

The notion of genetic value is problematic according to this field conservationist because it enacts too much control over both other animals and other people. As a discourse, genetic value reproduces the idea that humans can reproduce "wild" animals and in turn become "the god of the new creation." In other words, according to this conservationist, it reproduces rather than intervenes in the mastery discourse seen in efforts to transpose domestic and endangered animal bodies. This creates a patriarchal relationship, one in which the life sciences arbitrate how animals will live and how human-animal relationships will be forged. As such, this field conservationist believed that genetic value contains an ideology regarding the meaning of wild life that needs to be at the least opened up for debate because the genetic is here prioritized over the spatial and interactional. The field conservationist's goal is to resist patriarchy, a point made by paralleling women's fight for the right to control their bodies with previously colonized nations' rights to control how they will live with nature into the future as part of their heritage.

The worry here is that the genetic values of zoos risk becoming a hegemonic mode of interacting with wildlife, one that this field conservationist resists in order to sustain the centrality of space and place in species preservation. Charis Thompson (2005) has conceptualized the political economies of assisted reproduction through her notion of "the biomedical mode of reproduction." Drawing upon Karl Marx's ([1867] 1978) "mode of production" that delineated the ways in which social life is ordered according to different relations to production, Thompson asks how social life is reordered when reproduction is a central activity in capitalism. Within this conceptualization, Thompson addresses the question of alienation, wherein the risk of being alienated from one's labor expands to include the risk of being alienated from one's body parts. Thompson contends that commodification, which in the Marxist sense implies becoming a fetishized thing unto itself, is not the key problem in the biomedical mode of reproduction. Rather, the problem lies in issues of custodianship.

In the case of cloning endangered animals, body parts are at systematic risk of being alienated not only from individual animals and the spaces in which they reside, but also from the local groups charged with caring for those animals. Frozen zoos and kinship charts, as used in zoos, allow animals to be uncoupled from space in new ways. Exchanges in genetic material entail an exchange in who has the right to care for that material. As such, intervening in genetic exchanges becomes a site through which different ways of living with nature and animals are contested. Specifically, genetic definitions of endangered animals are being resisted and negotiated through the expansion and curtailment of endangered animal tissue economies.

The Biopolitics of Genetic Values: Domestication

Pluralist technology developers have been worried that the animals produced through interspecies nuclear transfer contain mitochondrial DNA from a domestic animal. In an attempt to avoid contaminating the gene pool of endangered animals, these scientists have only cloned male animals on the basis that their mitochondrial DNA will not spread to the population. The cloned animal has been "purified" (Latour 1993) by differentiating its biological parts (e.g., its genes) from its social parts (e.g., its mode of reproduction). What connects these two different parts is the assumption that genetics is the most real version of reality.[18] However, the focus on domestic animal mitochondrial DNA belies the extensive use of techniques with endangered animals that are associated with domestication. In this context, some have questioned what the use of techniques associated with domestication mean when used with purportedly "wild" animals. As the field conservationist I interviewed pointed out, it becomes difficult to distinguish the genetic management of endangered—and thus "wild"—animals from the centuries-old practice of genetically managing cats and dogs (Interview, April 25, 2006). If domestication has had such a great impact on other animal species, why should we assume that these practices will not also have an impact on endangered animals? Does the focus on mitochondrial DNA in cloning simply obscure a far wider set of changes occurring to supposedly wild animals within the zoo? This encapsulates the missing position discussed in chapter 1.

Broadly speaking, domestication is a form of interspecies relations wherein interdependencies are forged.[19] Beyond this broad definition, there are many ways of conceptualizing domestication more specifically. The most prominent understanding is, according to Barbara Noske (1997: 6) "the capture and taming by man of animals of a species with particular behavioral characteristics, their removal from their natural living area and breeding community, and their maintenance under controlled breeding conditions for profit." According to this familiar definition, domestication is purposeful human mastery that reconfigures the animal's body to fit the specifications of human desire, particularly profit, through a hierarchical relationship. In this context, domestic animals have come to be viewed as species that humans created (Ritvo 1997). This has served as an epistemic basis for understanding domestic animals as objects produced by humans, which can be legitimately owned, used, and consumed.

Zoo animals have long occupied an uncertain position in relation to "wild" and "domestic." On the one hand, animals living in zoos are not domesticated in a technical sense because most do not breed in captivity over a number of generations, despite sustained efforts.[20] Furthermore, even those animals bred in the zoo are not "tame."[21] On the other hand, metanarratives often position captive animals as symbols of human mastery in a manner that is continuous with predominant definitions of domestication.[22] While the wild animal is located in nature as free, the zoo animal born in the park seems to have more in common with domestic animals that have also been produced and kept as part of human culture. Keeping animals in captivity certainly shapes their social lives, behaviors, and biologies. Zoo animals thereby exist in the space between domestic and wild. "Wild life" as opposed to "wildlife" helps to mark out these differences.

As zoo animals have increasingly become protected ambassadors of their endangered species (Hanson 2002: 171), endangered species also occupy this borderland position. In her discussion of the relation between zoo and domestic animals, Kay Anderson (1998b) has argued that captive wild animals are domesticated by being brought into human concern and management. Drawing on Anderson's argument, it is possible to also posit that endangered species are increasingly being domesticated. However, the ostensible goal in forging a human relation with an endangered species is not based on captivity per se, but rather to help a population eventually become "self-sustaining" (Benirschke 1986) or viable without ongoing human intervention. Thus, humans relate to endangered animals through the practices associated with domestication, but the idea is that this relation should be conducted on a short-term basis and without long-standing genomic and behavioral consequences for the species in question. On this basis some have argued such animals should be understood as "protected," and therein as different from either domesticated or wild animals (Harris 1996).

Zoo and endangered animals are not the only kinds of animal that exist in these borderlands. Indeed, many animals are rather difficult to classify as either wild or domestic. For example, Marianne Lien (2007) has asked how and in what ways farmed salmon link up with and diverge from prominent definitions of domestic animals. Meanwhile, Yuka Suzuki (2007) has explored the production of wildlife in post-colonial Zimbabwe for tourism, a practice that contradicts traditional distinctions between wild and domestic. Not unlike zoo animals, both salmon and wildlife are being made through practices associated with domestication but nonetheless defy the category of "domestic."

In this context, recent scholarship has sought to reconsider domestication, no longer using the word to designate a kind of animal but rather using "domesticatory practices" in order to refer to a set of relational practices (Cassidy and Mullin 2007).[23] Shifting the terminology to "domesticatory practices" has allowed scholars from across the social and life sciences to ask how techniques like selective breeding have been used across a range of different species, and with what kinds of varying consequences. This research has emphasized that domesticatory practices can be engaged in through a hierarchical and domineering relationship, but can also be engaged in through a mutual relationship premised upon cooperation and exchange (Cassidy and Mullin 2007: 2–3).[24] Thus, just because SSP managers and pluralist technology developers are using domesticatory practices does not mean that they are seeking to control endangered zoo animals. Indeed, based on my research it seemed that pluralist technology development was far more rooted in the idea of shepherding, as opposed to mastering, endangered populations. The idea was that certain populations of endangered animals may be able to persist if humans take the time and care to shepherd their genetic body. Biological control was not the goal in this instance. Responsible practice was.

That said, shepherding animals is a constitutive practice. It is likely that a slew of traits are unconsciously selected for when one species shepherds the genetic configuration of another. Indeed, physical anthropologists have highlighted the salience of unconscious selection in domestication by pointing out that human bodies have been transformed through unintended selection pressures associated with domestication (Leach 2003, 2007). Humans share many of the physical characteristics commonly used to demarcate domesticated species, and have thus been domesticated through domesticating other animals. As such, domestication is increasingly being thought of as a two-way relationship that shapes both species involved, not only in social but also in biological terms (Haraway 2008). Vinciane Despret (2004: 122) has, in this context, reframed domestication from docility to an interaction wherein two different species make themselves available to the beliefs and concerns of the other in a manner that becomes mutually embodied. Humans and animals change in relating with one another through domesticatory practices. Engaging with endangered animals through cloning could very well change both species involved in unknown ways.

Redefining domestication in this way allows the biopolitics of zoos to be reconsidered. Matthew Chrulew (2011) has applied the concept of biopolitics to consider the practices of zoological parks over time. He has argued that the modern, colonial zoo embodied the biopolitics of the camp that Agamben (1998) has theorized. Animals living in small cages behind bars were "bare life" (Agamben 1998), in that these animals had come to stand for nothing but their anatomies, to be seen and represented by natural historians (Chrulew 2011: 143). Chrulew has contrasted this with the contemporary biopolitics of zoological parks, which he contends embody the productive facilitation of life that Foucault (1978) theorized. Selective breeding, well-researched enclosures, proper nutrition, and enrichment exercises are all meant to nurture and sustain the life of both zoo animals and captive populations to avoid the death of species. However, Chrulew has argued that zoo animal life is nonetheless "wounded life" within this biopolitical regime because of the totalizing control that zoos exert. Shepherding and control are the same in Chrulew's analysis.

Contemporary scholarship that has reconsidered domestication can, however, modify this analysis. Specifically, Chrulew references the zoo animal in general across his analysis as the subject of totalizing control; he does not reference the ways in which techniques are used with and adapted to specific kinds of zoo animals. In contrast, the idea of domesticatory practices shows that a set of techniques can be used in different ways with different animals. This book demonstrates this point by showing how cloning is used differently with not only endangered when compared to domestic animals, but also among varying endangered animal cloning projects. In this context, domesticatory practices in and of themselves are not bad, or even necessarily controlling. Rather, the important question is whether and how humans and other animals are able to "make themselves available to the concerns of one another" (Despret 2008) within the space of the zoo and with the requisite domesticatory practices. In this context, it seems probable that the life of some species can flourish within these relations, whereas the life of other species cannot. For example, it is highly unlikely that humans can make themselves available to the concerns of the polar bear within the confines of a zoo, as this species can swim up to five hundred miles in a single day (Clubb and Mason 2003). On the other hand, there are good reasons to believe that humans and banteng can make themselves available to the concerns of one another within zoological parks. There is, after all, a semidomesticated herd of banteng in Australia (World Association of Zoos and Aquariums).

The human is thus implicated in the cloned animal in a rather different way here. Rather than control the zoo animal through cloning, humans are instead forging relationships with endangered animals through domesticatory practices that are constitutive of both species. There may be a whole slew of unseen processes and unintended consequences that therein result for both species involved. After all, domestication started as "unconscious" forms of selection that developed in reaction to new environments and cultural practices (Leach 2007). And Vinciane Despret (2004) has shown that some animals, such as horses, are particularly adept at reading human bodies in ways that can shape what both humans and horses do with their bodies in interaction. It thus becomes possible that humans are changing themselves and endangered animals in and through their domesticatory practices, which includes cloning. Biopolitics here denotes less the ways in which human and animal management regimes are interrelated and coconstituted as seen in transbiology, and more of the ways in which humans and animals live and become lively (or don't) in and through regimes through which different species must necessarily relate.

Preserving Wild Life

Amongst the trees and picnic benches of Griffith Park in Los Angeles, one can visit the city's "old zoo."[25] Standing as a kind of archaeological site for zoo historians, the small, barred cages of this former zoo have been kept for prosperity. Deemed improper accommodations, the zoo animals were moved out and a host of other species have since made the cages their own. Rather than display animals, the old zoo instead stands as a reminder of what the zoo used to be like and a signal of how much this institution has changed.

Illustration 5.1: The old zoo in Griffith Park, Los Angeles, Calif.
(Photograph taken by Stephanie Miller; reprinted with permission.)

About a mile away is the "new zoo" of Los Angeles. This zoological park looks like other zoos I have visited, which many readers are likely familiar with. Here animals are kept in larger spaces and are housed with other animals. Moats often keep zoo animals within their space, and cages are visually minimized to create an appearance of freedom.[26] Zoo visitors see animals whose species are diminishing in an environment that is meant to look like their original habitat. Nature is something that needs to be seen and appreciated so that its "real" counterpart will continue to be protected. I overheard one zookeeper summarize this to a group of schoolchildren as follows: "Whose fault is it [that these species are endangered]? People. Who is trying to help them? People. So you have the people who hunt them and the people who help them." The zoo stands as nature's helper in this context.

Gregg Mitman has interpreted the shifts in zoo animal display practices embodied by the old and new zoos in Los Angeles. His essay entitled "When Nature Is the Zoo" (1996) starts with the African Plains exhibit at the Bronx Zoo, which was launched in the 1940s as the first immersion exhibit within the United States. Animals were not organized according to taxonomic classifications here, but rather by continent and habitat. The goal was to "re-create nature" as part of public education. Mitman notes that the exhibit planner, Fairfield Osborn, eschewed both the idea of technological control over nature as well as the separation of humans from nature. Osborn's goal was to instead "aid and abet"—but not replace—natural processes. In this context, Osborn wanted to create a new way of seeing animals in the zoo that would mimic the sublime experience of seeing animals in the wild. However, this necessitated that "any trace of artificiality, any visible manifestation of power, had to be rendered invisible" (Mitman 1996: 120). Mitman has emphasized that visual technologies made these dual goals possible.

If Mitman was interested in the changing display practices of zoological parks, this chapter has explored the zoo's changing collection practices. Where Osborn wanted zoo visitors to see wild animals in new ways, the scientists presented in this chapter have wanted zoos to collect wild animals in new ways. This has required that humans control zoo animal reproduction. However, almost all the people involved in such practices shared Osborn's goal to "aid and abet"—but not replace—natural processes involved in animal reproduction. Nature has served as a guide in this attempt to use cloning in order to aid species threatened by a human-dominated planet.

But in stark contrast to the visual practices described by Mitman, assisted reproductive technologies do not conceal themselves in the way visual technologies may. As Marilyn Strathern pointed out (1992a: 22), assisting biological reproduction makes it rather difficult to think of biology as separate from social life. Reproductive technologies do not hide from view the entwined and dynamic relationships between nature and culture, biology and technology, wild and domestic, as well as animal and human. Pluralist technology development, pursued through the logic of genetic value, has sought to delimit the biological changes made to endangered animal bodies by delimiting the presence of mitochondrial DNA. None of the supposedly "social" practices involved in managing captive endangered animals are deemed constitutive by privileging sexual reproduction and heredity. But there has, nonetheless, been a strong counterargument that rejects this "purification" (Latour 1993) practice. The biological and social elements of endangered animals seem to "loop" (Hacking 1999) and thus change one another, rather than be discrete.

In response, some have collapsed the transbiological goals of technology developers with the genetic values of population managers. Both are rooted in making rather than finding endangered animals and populations. Both require human control over the genetic configuration of endangered species. Lumping these projects together usefully shows these continuities, and thus highlights the role of human activity in delineating the bodies of endangered animals within zoos. Lumping these projects together also resists the strategic essentialism of genetics in zoos. The concern is that this valuation system risks eclipsing other ways of living with nonhuman animals.

However, there are also important differences between transbiology and genetic value as social and biological projects. Splitting these projects apart allows us to see that endangered species are nonetheless constituted in different ways across these projects. Transbiology hopes to redistribute endangered animal reproduction so that humans can control these processes through technology. There is a technophilia undergirding this approach. Meanwhile, all the population managers I spoke with emphasized that replacing natural reproduction is probably impossible because it would be highly inefficient and cost far too much. They were keenly aware of the limits of technologies, and how little humans can control biology when it comes to endangered animal reproduction. Cloning has not been pursued within the logic of genetic value for the purpose of technological mastery. The goal has instead been to engage in responsible relations with endangered animals by guiding the genetic configuration of a population. Genetics provides a means for zoos to pragmatically pursue the zoo in a new and more ethical way. Rather than remake endangered animals, the goal here is to remake the zoo itself.

What kind of "wild life" is enabled by the genetic values of zoological parks? To answer this question, it is useful to remember Paul Wapner's (2010: 155) point that humans cannot determine the animals that they create. The banteng is other to humans, and can resist what humans may want him to do. Like many other zoo animals, on a very basic level the banteng might resist the zoo by not producing offspring. So the banteng represents an otherness that is nonetheless embroiled with humans. But he nonetheless does not fully embody "wild life" as delineated by Clark (1997, 1999). The banteng was not pursued so that his otherness would break free from and challenge the biological order of the banteng species. His otherness was in fact actively contained by his maleness. He was not produced out of a fascination with the unruliness of biology. Rather, the banteng was produced because he is other, and because zoos must be responsible to other species in an unruly world.

However, one of the unintended consequences of the banteng is that, while he has been pursued through the logic of genetic value, he has also raised questions about its limits. It becomes clear that the environment is at best strategically marginalized in this articulation of cloning. This forces us to ask whether cloning could be articulated with the questions and concerns of those interested in space and place. What might such an articulation look like? The next two chapters will address this question.

6

Knowing Endangered Species

A number of people I spoke with while conducting this research were rather critical of endeavors to clone endangered animals. All affiliated with zoos, these scientists questioned if animals produced through interspecies nuclear transfer counted as "endangered" due to their mitochondrial DNA being inherited from another, domesticated animal species. Many wanted to know more about interspecies mitochondrial inheritance, its role in embryonic development, and its consequences for the health of offspring. In other words, these scientists wanted to take a precautionary approach to biotechnology development in the zoo.[1] According to this perspective, more scientific knowledge is required regarding both normal embryonic development in endangered animals and the influence of nuclear transfer on that process before the risks associated with cloning can be tolerated.

In this context, and not surprisingly, many critics decided not to engage in cloning research. For example, one zoo scientist was somewhat upset by my even researching cloning in zoos. In an email, she explained why she and her colleagues do not think cloning endangered animals is a justified scientific project. When techniques like artificial insemination and embryo transfer have not yet made any substantial contribution to species preservation, they thought that cloning was premature. They were also concerned about the consequences of producing heteroplasmic animals through interspecies nuclear transfer. She concluded by stating that the most important thing in species preservation is diversity, and cloning presumably cannot assist with this effort. Any cloning research was therefore deemed illegitimate.

In contrast, three people I spoke with who were critical of cloning endangered animals had nonetheless become involved in cloning projects. Bill Swanson has been a vocal critic of cloning endangered animals, but nonetheless received funding from the National Institutes of Health to conduct domestic cat cloning experiments in collaboration with a private pharmaceutical company.[2] Barbara Durrant has been concerned about cloning zoo animals, but told me that she has nonetheless learned a lot from biotechnology companies involved in cloning domestic dogs (Interview, May 16, 2006). That said, neither Swanson nor Durrant articulated the benefit of these cloning projects in terms of producing a cloned endangered cat or dog. Rather, the domestic cat cloning project brought research funds into Swanson's lab and was linked with his use of domestic cats as "surrogate models" (Bolker 2009) of endangered felines. Similarly, Durrant uses domestic dogs as models of endangered bears. Because more basic knowledge regarding the reproductive physiology of dogs had to be developed in order to use somatic cell nuclear transfer, dogs consequently serve as better model organisms. In both these instances, cloning facilitated the creation of better knowledge regarding the reproductive physiologies of domestic animals that in turn serve as indices of endangered species. Cloning was not used as a reproductive technology per se here, but rather in conjunction with the animal model paradigm.

This chapter focuses on how cloning is being articulated as basic reproductive science research in zoos through the logic of modeling. This articulation of cloning has to date been most fully elaborated in a research project at the Zoological Society of London, which has been focused on amphibian cloning. After describing how and why frogs are being cloned in this context, I consider how this articulation of cloning refracts the turn toward more basic scientific questions in the reproductive science research programs of some U.S. zoos. The chapter concludes by asking how nature and culture are linked in this articulation of cloning, wherein the constitutive role of knowledge practices is highlighted in rendering nature a meaningful entity and site of action.

Rearticulating Cloning

While I was conducting the initial research for this book, a number of people who were critical of cloning suggested that I interview Bill Holt, a reproductive scientist at the Institute of Zoology, Zoological Society of London. People told me that he had been a key figure in outlining why cloning endangered animals is premature and problematic. So, upon moving to London in 2009, I contacted Holt and asked if we could meet. He kindly agreed and we met for an informal conversation shortly thereafter.

I met Holt at his office in the Institute of Zoology, which is across the street from the London Zoo in Regent's Park. Meeting me in the lobby, Holt led me through a set of locked doors, followed by a series of narrow and slightly darkened corridors. Walking through these hallways, I thought that the Institute of Zoological looked and felt much more like the university laboratories I had visited. This contrasted with the more manicured research facilities of most zoological parks I had seen in the United States. The built environment at the Zoological Society of London seemed to embody the basic scientific approach that so many people in U.S. zoos were arguing for, the kind of work Holt was known to do.

Upon arriving at Holt's office, I explained my research and told him that a number of people in the United States had suggested we speak. I continued by asking if he would be willing to tell me how he came to be involved in the debates over cloning endangered animals and why he was seen as a spokesperson for people who had reservations about these experiments. Holt smiled humbly. He quietly stated that his American colleagues were very generous, but he would certainly tell me about his experiences with cloning endangered animals.

Holt began by recounting the extensive—and possibly excessive— enthusiasm for cloning endangered animals that arose shortly after the birth of Dolly the Sheep, at the annual meetings of the International Embryo Transfer Society. He noticed that this excitement was largely articulated by people working in biotechnology companies, as opposed to zoos and species preservation. In this context, Holt noticed that a lot of effort was being directed at learning how to clone wild and endangered animals—specifically mammals. However, these efforts seemed to serve a publicity function as opposed to contributing to conservation goals. In this context, Holt began to articulate his reservations about cloning endangered animals. And as a consequence he was asked to participate in a debate on cloning endangered animals by taking the "anti"-cloning position. He was also asked to write a review outlining the practical potential of cloning for species preservation.[3]

In an interview that was conducted later, Holt described how he became involved in cloning endangered animals as follows:

> Going back, probably to the time of Dolly the Sheep being announced, I'm in a society called the International Embryo Transfer Society.... And I was quite surprised by the level of, what I thought was, unthinking enthusiasm. [People were saying,] "Now, okay, Dolly the Sheep has been produced; therefore we can clone more or less anything we like because the techniques are around." . . . I felt a real [sense of] misery because I was saying, "No, you should hold on because it's difficult even to do artificial insemination." . . . If you go up the technological scale to embryo transfer . . . if you go through the literature, the number of successes with that technique tends to be mostly one offs. And that partly has to do with this idea of "okay, we'll do it and if it's successful then we'll get the press headlines and then we don't need to do it again." But there's a lot of effort that goes into getting all those things right. So to go to the next step and clone when you know from the Dolly experience that the success rate of just producing embryos is really low. And you're working with wild animals where you know very little about the basic reproductive biology.... So that's why I was being a bit cautious in saying, "Look, hold on. There are too many unknown things when you work with wild species."
> *Interview (May 20, 2010)*

Here, Holt outlined his belief that cloning was unlikely to be an effective tool for conservation given the difficulties with other reproductive technologies that are easier to use. These problems arise because there is insufficient basic knowledge regarding the reproductive biology of most endangered species. These problems also arise because so much effort is put into creating a spectacular animal through sensational reproductive technologies so as to get media attention, rather than spending time making the technique efficient enough to use in ex situ species preservation. In this context, Holt became a spokesperson for taking a more precautionary approach to developing biotechnologies like cloning.

Holt paused at this point in the conversation. Many of Holt's concerns regarding cloning had become familiar to me by this point, and so I was about to take advantage of this pause to ask a follow-up question that might open up some new aspect of these concerns. However, before I could articulate such a question, Holt continued in a manner that I could not have anticipated. "So," he said, "it might then come as a surprise to you to learn that I am currently involved in initiating a new cloning project." I was without a doubt surprised, and very eager to find out how and why one of the world's leading critics of cloning endangered animals had decided to bring nuclear transfer home to the zoo.

Holt went on to explain that Rhiannon Lloyd, a postdoctoral fellow at the Institute, had received a grant through the Leverhulme Trust to conduct cloning research with amphibians under his supervision.[4] Lloyd's research examines the relationship between mitochondrial and nuclear DNA in cellular development, using interspecies nuclear transfer with two common species of frog. Here, molecular biology—as opposed to a live birth of offspring—arbitrates questions about the feasibility of interspecies nuclear transfer. The question here is not "Can we produce an offspring using interspecies nuclear transfer?" The question is instead: "Do the problems associated with nuclear transfer arise because of incompatibilities in the mitochondrial and nuclear DNA that compromise developmental processes?"

In answering this question, Lloyd is learning how to clone frogs in a manner that parallels the work of scientists at ACRES. If nuclear transplantation is considered biologically feasible and is made efficient enough with amphibians, it may be possible for her to clone endangered frogs as part of species recovery programs in the future. And because frogs produce a large number of eggs outside their bodies, endangered amphibian cloning should not require the interspecies modification that has marked all cloned endangered animals to date.[5] By cloning amphibians, Holt and Lloyd are actively carving out an alternative articulation of cloning, one that seeks to conduct basic physiological and genomic research in the zoo that is interlinked with species recovery programs. They are also articulating cloning in a way that does not introduce genetic changes to the target species.

Cloning Amphibians as Basic Science

Lloyd's research is supported by the Leverhulme Trust, which funds basic scientific research. In this context, the questions she asks are centrally focused on basic biological processes as opposed to conservation applications. Holt commented during an interview: "We're applying to a scientific body. If you write that you want to do cloning for conservation, they'll immediately throw it out. So you find scientific questions that are valid and then you address those questions. And the conservation is a bit on the side" (Interview, May 20, 2010). Not unlike the cloning projects involving biotechnology companies, this research project is not centrally organized around a conservation priority. In both instances, conservation has been described by involved scientists as being "a bit on the side." But where biotechnology companies and ACRES have used cloned endangered animals as proofs of the interspecies nuclear transfer concept, Holt and Lloyd are instead using interspecies nuclear transfer as a model system for asking questions about embryonic development. Rather than make an animal, they instead hope to better understand the role of mitochondrial DNA in development.

Using frogs to ask basic scientific questions about cellular development symbolizes the ways in which this research is part of basic science, as frogs have played a key role in experimental biology.[6] Indeed, both cloning and frogs have been central to experimental physiology across the twentieth century. Jane Maienschein (2001, 2002, 2003) has historicized cloning in this context, and notes that Hans Spemann is considered the initiator of the nuclear transfer technique, having first inscribed cloning as a potential means through which questions about embryonic development could be answered. Spemann used experimental techniques, largely with frogs, to answer questions about cell differentiation. He was curious about what would happen if the nuclei of cells were transferred to enucleated eggs at different stages of development (Maienschein 2003: 115–116). By changing reproductive processes in this way, Spemann thought that much could be inferred regarding normal reproduction.

In 1952 Robert Briggs and Thomas J. King implemented cloning in this way in order to ask if individual cells differentiate as different parts of the embryo differentiate. Briggs and King took nuclei from frog eggs of one species and put them into frog eggs of the same species and of another species. While Briggs and King were able to get nuclei from eggs in the early stages of division to cleave and develop, they were unable to get later stage transplants to do so. They concluded that the nuclei of cells in later stages of development were not able to reprogram and cleave because irreversible changes occur when a cell differentiates (Maienschein 2003: 118–120). However, in the 1960s John Gurdon announced that he had been able to reprogram a differentiated frog cell, challenging the notion that irreversible changes occur in the nucleus of amphibian cells (Maienschein 2003: 122–123).[7] In interviews, both Holt and Lloyd located their research vis-à-vis Sir John Gurdon's amphibian cloning research. Indeed, Lloyd learned how to clone in Gurdon's lab. As such, their research is embedded, both symbolically and in practice, within a long history of asking basic questions regarding embryonic development using both cloning and frogs.

As I write in September 2010, Lloyd is learning to clone using a common frog species, Xenopus laevis, that is frequently used in scientific research. She is trying to perfect her cloning technique at the European Xenopus Resource Center in Portsmouth, United Kingdom. Not unlike researchers at ACRES, she is struggling to become skilled at nuclear transplantation. Lloyd explained in an interview:

> Currently I'm only getting up to the morula stage, where the embryo basically tries to take on everything rather than the maternal genome. And they're failing at the moment. I think it's a technical thing, rather than a nucleo-mitochondrial genome thing. I haven't quite got the right apparatus. . . . So it's a bit of a challenge. It's not as easy as I think people think, at least not in my hands. Maybe I'm particularly bad at it. Because I think Bill thought it would be really easy to do. People always say "oh, why are you doing that? It was done in the '50s. Surely it's really easy." And perhaps if you're John Gurdon and you've got all this experience, perhaps it is really easy. . . . I think I thought it was easy before I started doing it as well.
> *Interview (June 16, 2010)*

Just as ACRES had to perfect their technique before producing offspring—by bringing cloning home to their hands in the zoo—Lloyd must also perfect her technique before addressing questions about nuclear-mitochondrial DNA interactions in embryonic development. Once her cloning apparatus is more fully established, she plans to conduct initial research using interspecies nuclear transplantation with Xenopus laevis and another common frog species, Xenopus tropicalis. Through this research, Lloyd hopes to better understand the role of mitochondrial DNA in embryonic development generally, as the frog has long stood as an "exemplary model" (Bolker 2009) whose biology can serve as an index from which a full range of different species can be compared. In the process, Lloyd and Holt also hope to determine if nuclear transplantation can be used with endangered frogs as part of species preservation efforts.

Cloning Amphibians as Conservation

By conducting cloning experiments with frogs in order to answer basic questions about cellular development, Holt and Lloyd have articulated cloning as basic scientific research. But does this research similarly come at the expense of species preservation? How will learning about mitochondrial-nuclear interactions in the cellular development of common frogs help conservation efforts? To answer these questions, it is important to emphasize that frogs resonate symbolically in both scientific and conservation communities. I will now discuss how Holt and Lloyd have interlinked their cloning work with endangered amphibian preservation, not only symbolically but also in practice.

It needs to be emphasized that frogs are an important species within the conservation community, as populations are declining at an accelerated rate worldwide. As a taxa, amphibians are considered more threatened than either mammals or birds (Beebee and Griffiths 2005; Stuart et al. 2004).[8] An Amphibian Action Plan (Gascon et al. 2005) was devised in response, which brings multiple components of conservation together, including habitat surveillance, captive breeding, and cryopreservation.[9] An Amphibian Ark (AArk) was created in this context. Through a partnership between the World Association of Zoos and Aquariums, the Conservation Breeding Specialist Group, and the Amphibian Specialist Group, the goal of AArk is to breed endangered amphibians in captivity. Cryopreservation plays an important role within this mandate, in which Holt, Lloyd, and Oliver Ryder are all actively involved. The AArk hopes to expand its existing collections of amphibian egg cells at key centers across the world, including the San Diego Zoo Global and the Natural History Museum in London.[10]

Endangered amphibian conservation efforts operate somewhat differently than the Species Survival Plans described in chapters 4 and 5. Many amphibian populations are at serious risk of extinction, which means that the number of individuals within the population is extremely small. The reasons for population decline are not generally understood. Where it is well known that habitat destruction is the key reason for declining numbers of banteng, many of the habitats of endangered amphibians appear to be intact (Stuart et al. 2004; Beebee and Griffiths 2005; Gascon et al. 2005). In this context, conservationists are at times taking the entire population of endangered amphibians into captivity before the population goes extinct.[11] The goal is to stabilize the population by expanding its numbers through genetic management. Meanwhile, field conservationists determine what environmental factors led to the initial decline and work to rectify the habitat. When both the captive population and the in situ habitat have been stabilized, the amphibians are expected to be reintroduced into their native habitat.[12]

Importantly, these endangered amphibians are not collected by zoos for the purpose of display. And their reproduction is not managed vis-à-vis an in situ population. Rather, these small amphibian populations are a closed population held entirely in captivity due to the severity of their endangerment. Holt explained this kind of species recovery program to me as follows, and the potential role of cloning within it:

> Zoos are taking it upon themselves, if they think a population is endangered, to go and capture the entire population and take them into captivity. They set up a biosecure facility specifically to manage that population. So, I mean, that population therefore then doesn't exist anywhere else. So, you're almost bound to lose genetic diversity. If you could clone, you could possibly regain—in the same way as the banteng—that diversity, and put it back. The same would apply if you had frozen sperm. So that whole suite of techniques would have a place in trying to maintain the diversity of that population. Who knows how long these populations have to be kept [before reintroduction]. It might be a hundred years or more, and if you've got banks of tissue and sperm you can do your best to keep that population viable.
>
> *Interview (May 20, 2010)*

The way in which Holt articulates the significance of cloning in the above statement resonates with another species recovery program, that of the black-footed ferret. This species had been presumed extinct. However, a farmer's dog found a black-footed ferret in Wyoming during the early 1980s. In response, the U.S. Fish and Wildlife Service collected all black-footed ferrets in the region, which was ultimately less than thirty individuals. The goal was to stabilize the population in captivity and then reintroduce the animals into their habitat. In this context, it was extremely important that each individual reproduce in order to ensure that the population grew in numbers without becoming inbred. However, without knowledge of how to raise black-footed ferrets in captivity, a number of animals died.

In this context, reproductive scientists at Conservation and Science, a Smithsonian Institute research center affiliated with the National Zoo, argued that artificial insemination could provide a useful service, in that deceased males could still reproduce using cryopreserved sperm and artificial insemination.[13] Budhan Pukazhenthi, a geneticist at Conservation and Science, described the research center's role in the preservation project as follows:

> Nobody knew how to raise black-footed ferrets in captivity. So what happened over the next few months or couple of years was they lost a few animals. So you have the last 28 animals and you lose 6 or 10 of those. That is not in the best case scenario. . . . Then researchers—including our institution—jumped on board. We expressed interest in developing artificial insemination so that we can at least ensure a fair representation of all the animals in the existing population. So over the years, almost twenty years now, the population has been stabilized in captivity. And the population has done relatively well, to the point that they're having animals reintroduced to where they had originally been present. . . . We have reintroduced close to 100 or 150 animals through artificial reproduction and over 500 animals are back in the wild now. So it is a pretty impressive success story. But it also took a lot of time. This story could have gone in the opposite direction if we had not managed to save some of these animals that we did with artificial insemination.
>
> *Interview (April 11, 2006)*

For many scientists I spoke with, the black-footed ferret represented an exemplary model of what reproductive scientists working in zoos should try to achieve. Here basic knowledge regarding reproductive physiology was developed with a priority species in order to develop a simple assisted reproductive technique (e.g., artificial insemination). Knowledge production and technology development are here part of a concerted preservation project, which is akin to the pluralist technology development seen with the banteng.

Holt and Lloyd are working to interlink their cloning research with ongoing efforts to preserve endangered amphibians. For example, Lloyd played an instrumental role in organizing a workshop in September 2010 that brought basic scientists working with amphibians together with amphibian conservationists. The workshop took place at the Zoological Society of London and the European Xenopus Resource Center in Portsmouth. The goal was to allow biologists and conservationists to see each other's work, and to collaboratively consider how they could help each other in conserving endangered amphibians. This workshop served a similar function as Oliver Ryder and Philip Damiani's meeting with the Taxonomic Advisory Group for endangered hoofstock, in which they explained how the banteng would assist with the genetic management of the zoo population. However, this workshop also extended the pluralist goals of the banteng cloning project. It occurred earlier in the cloning research endeavor, which allowed Lloyd the opportunity to incorporate a range of people's concerns into her cloning experimentation from the start rather than informing people of how her research could assist them in their preservation efforts at the end. A core argument of science and technology studies has been that the social, ethical, and legal discussions about technoscience should occur "upstream" in the development process, as certain priorities come to be embedded in scientific and technological development at these early stages that are difficult to change later on.[14] Lloyd's work embodies such an approach to reckoning with the politics of contemporary cloning practices in the context of species preservation.

Basic Science in the Park

As already discussed, the reproductive sciences entered U.S. zoos through technology development. However, some reproductive scientists I interviewed were rethinking this research priority and were reorienting their work toward the basic scientific approaches that are embodied by the amphibian cloning project. Some now focus on making low-tech reproductive technologies more efficient with endangered animals. Others ask more basic questions about the physiologies of unstudied, endangered species.

This shift from technology development to basic science is most vividly demonstrated in David Wildt's (2004) discussion of his own research trajectory, which began in the early 1980s with the goal of developing artificial insemination with cheetahs. Consistent failures led Wildt to what he has called his "ah-hah" moment, wherein he realized that "a cheetah is not a cow" (Interview, July 18, 2006).[15] Here, Wildt realized that the techniques and protocols used to reproduce domestic cows would not necessarily work to reproduce a cheetah. In response, he and his colleagues began a fifteen-year study of the cheetah's reproductive physiology.

Wildt has since shifted his scientific practices so that he starts with basic, biological questions as opposed to technology development. He articulated this focus as follows:

[A]ll these species have these unique physiological characteristics and that itself dictates whether or not a particular technique has any relevance. So the bottom line is that, for me, the technology—whether we're talking about semen collection and sperm evaluation or we're talking about cloning—the value of that technology is really as a tool to understand species uniqueness, the different mechanisms among species that make them so different from one another and allow them to be reproductively successful and able to transfer their genes from generation to generation. It's not—actually, it has very little to do with the ability to produce offspring.

Interview (July 18, 2006)

Here we see the idea that basic science should not be evaluated according to the birth of an animal. Scientific practices are instead evaluated by asking if there has been an increased understanding regarding the diversity of biological forms and functions. The spectacle is not an animal produced through technoscience here, but rather the ways in which technoscience can illuminate the spectacular uniqueness of different species.

That said, basic knowledge can lead to the refinement of existing techniques for endangered animal reproduction, such as the eventual use of artificial insemination with cheetahs. Or, basic research can lead to the creation of new kinds of technologies based on interaction with endangered rather than domesticated animals. Wildt continued:

> One of our most recent high-profile accomplishments is the production of Tai Shan, the giant panda cub. That animal was produced not using cow technology or human technology, but was from a database that we had developed over the course of about nine years working in China with our Chinese colleagues, worrying about Giant panda sperm and how to collect it, how to keep it alive, how to put it in a female at the right time, which is totally different than what you might expect for a cow.
> *Interview (July 18, 2006)*

Here Wildt implicitly critiques the logic of transposition that much of technology development has been premised upon. Instead, he argues that endangered species should be the focus of research. This produces basic physiological knowledge, which can be the basis for developing new technologies with and for endangered animals as opposed to transposing techniques produced for domestic animal reproduction.

As previously mentioned, Adele Clarke (1995) has argued that there have been two phases in U.S reproductive sciences. Modern reproduction focused on the structures and functions of reproductive organs in order to control reproductive processes as part of family planning, while postmodern reproduction has sought to transform those reproductive processes for the purpose of identity creation.[16] Drawing on Clarke's schematization of the reproductive sciences, I posited earlier that U.S. zoos began to engage in the reproductive sciences during the rise of postmodern reproduction, with the hope that they could bypass modern reproductive research on zoo animals and jump right into postmodern reproductive transformations. However, this research trajectory has increasingly been called into question, as some researchers are returning to a set of concerns more clearly aligned with modern reproduction. This includes developing basic physiological knowledge regarding endangered species as well as more simple reproductive techniques such as pregnancy tests.[17] The goal here is not to transform reproduction. The goal is instead

to understand how reproduction works across a range of different species. That understanding is considered useful in and of itself, but could
be instrumental in managing endangered animal reproduction. As such,
cloning can be articulated as either modern or postmodern reproduction,
and can be used as part of technology development or basic research. The
use of this biotechnology does not necessitate a particular approach to
scientific practice. It is flexible enough at present to accommodate a number of different ways to make nature through zoo science.

Modeling Difference

In the larger scientific field, interspecies nuclear transfer is bound up
in the shifting terrain of animal modeling, wherein species differences
are becoming an explicit object of inquiry. A review of interspecies
nuclear transfer points out that the development and use of this technique embodies an assumption that biological forms and functions
are generally conserved across different species as part of evolution
(Beyhan, Iager, and Cibellie 2007). The authors state: "The two main
assumptions required for iSCNT are that early developmental events
and mechanisms are evolutionarily conserved among mammals and
that molecules that regulate these events in mammalian oocytes are
capable of interacting with nuclei from another species" (Beyhan, Iager,
and Cibellie 2007: 503). This statement shows that interspecies nuclear
transfer requires developmental events and mechanisms, along with the
molecules regulating those events, to be more or less the same across
species. However, the authors continue that this assumption requires
validation, and cannot be presumed: "The validity of these assumptions, however, deserves vigorous scrutiny. Although most mammalian
embryos follow a very similar pattern of ontogenic development, significant differences in many aspects of the process do exist among evolutionarily divergent taxonomic groups. . . . Temporal regulation of developmental events—such as cell-cycle progression, embryonic genome
activation (EGA), blastocyst formation, implantation, and organogen-
esis—differ from species to species" (Beyhan, Iager, and Cibellie 2007:
503). The authors point out here that there are biological differences in
developmental processes across species. And these differences may be a
barrier to interspecies nuclear transfer itself.

In this context, the authors use their review to bemoan the lack of reports that examine the interactions and compatibility between the nucleus and the cytoplasm along with the fetal-maternal interactions in the context of interspecies nuclear transfer experimentation (Beyhan, Iager, and Cibellie 2007: 506). Indeed, the authors ultimately contend that interspecies nuclear transfer may be most useful as a model for exploring these biological processes across species. This is precisely what Lloyd and Holt's research at the Zoological Society of London seeks to address. They are using interspecies nuclear transfer as a model system for asking questions about the interactions between the nucleus and cytoplasm in development.[18] Rather than ask what role mitochondrial DNA plays in defining a species (as seen in chapter 1), Lloyd and Holt instead ask what mitochondrial DNA does in cellular development within and across species. In other words, they are looking at the ways in which genes are expressed in relationship with the cellular environment.

The shift toward using cloning to answer basic scientific questions regarding the differences between species is not only significant for reproductive scientists working in the zoo. The difficulties experienced while trying to transpose techniques and bodies from agricultural to zoo animals have epistemological consequences for the life sciences more generally. The animal model paradigm holds that species conserve physiological forms and biological processes through evolution. The importance of comparing different species is often implicit.[19] The frog has, for example, long been used as an "exemplary model" (Bolker and Raff 1997) in the life sciences, wherein findings are meant to stand for a full range of different species. However, the problems associated with transposing the bodies and technologies of domestication into endangered animal reproduction fissure the assumption that frogs can stand for a whole range of different species. Thus, what Holt and Lloyd find out about common frogs through their research will have to be compared and contrasted with endangered frog species. Comparison has thus been made explicit in the animal modeling practices of zoos.

I consider here how basic scientists in zoos have built upon discursive traditions within both the life sciences and conservation to argue for more physiological knowledge regarding a greater range of species. In doing so, they offer a critique of not only technology development in the zoological park but of the animal model paradigm in the life sciences more generally. By calling for more physiological knowledge regarding a fuller range of species, zoo scientists are resisting a core reduction that has long structured life science research. In turn, these scientists are reasserting the role of the zoo in the biosciences.

Beyond the Animal Model

In a review of ten reproductive biology journals, David Wildt (2004) found that more than 90 percent of text was devoted to fourteen species, including humans, cows, pigs, and mice. Based on these findings, Wildt (2004: 284) has concluded that "experimental biology is disproportionately directed at already well-studied species." Historian Harriet Ritvo (1995: 422) similarly found that domesticated animals occupied "more than their share of space" in the natural histories of taxonomists over two centuries ago, in the late eighteenth century.[20] Wildt has suggested that this excessive focus may be due to financial and logistical troubles of studying other species, alongside a general lack of initiative on the part of researchers to study other kinds of animals. Ritvo has come to a similar conclusion, arguing that domestic animals were overrepresented due to their economic importance and accessibility. I would add that the animal model paradigm also helps to explain why a select few species are so well studied.[21]

The solidification of the animal models paradigm across the twentieth century has been interrelated with the presumption that species conserve biological forms and processes through evolution, making it possible to know an array of species while studying only a select few.[22] Historian of science Cheryl Logan (1999, 2001, 2002, 2005) has explored when and how the presumption that species conserve biological forms came into being. Through a number of case studies, Logan has shown that the extent of physiological similarity and difference across species was not assumed at the end of the nineteenth century, but was instead an empirical question. It was not until after 1900 that species similarity became an a priori assumption in life science research. In this context, some species became "exemplary models" (Bolker and Raff 1997), whose bodies could stand as universal representations of biological forms and processes across a full range of other species. Logan (2001,

2002) attributes this shift to the increasing availability of standardized animals as research materials through commodification.[23]

Wildt has been critical of the small number of species studied as a consequence of this approach to physiological knowledge. He thinks that more species need to be known from a physiological perspective. This is especially true of endangered species, which Wildt believes must be understood if they are to be saved. And Wildt is not alone in his concern. Logan (2005) notes that the presumption of species generality has begun to fissure in the context of neuroscience. Meanwhile, Jessica A. Bolker and Rudolf A. Raff (1997) have similarly argued that presumptions of species generality are waning in development biology. As such, the zoo is another institutional site where presumptions of species generality are beginning to fissure in response to everyday scientific practices.

But in addition, Logan (2002) has pointed out that the presumption of species generality was resisted at different moments across the twentieth century.[24] She notes, for example, that in his opening address to the Thirteenth International Physiological Congress, August Krogh (1929) set out a vision for physiology that resisted the presumption of species generality. To stake this claim, Krogh provided an anecdote from Christian Bohr's research experience. According to Krogh (1929: 202), Bohr was interested in the respiratory mechanism of lungs, which he sought to study through a method that considered exchange through each lung separately. To conduct this experiment, Bohr found a species of tortoise whose trachea was perfectly designed for the experiment, to the extent that laboratory workers joked that the "animal had been created expressly for the purposes of respiration physiology" (Krogh 1929: 202). Krogh used this anecdote to conclude that there are many different animals that could be the "right" species for experiments, but they are not known to physiologists.[25] To produce better physiological knowledge, Krogh argued for an expansion of physiological knowledge in zoology and for greater cooperation across these two disciplines.

Krogh also located the need to understand a greater array of species in simple curiosity and fascination, a key if underappreciated feature of much scientific work. He stated: "I want to say a word for the study of comparative physiology also for its own sake. You will find in lower animals mechanisms and adaptations of exquisite beauty and the most surprising character, and I think nothing can be more fascinating than the senses and instincts of insects as revealed by the modern investigations" (Krogh 1929). Here we see the idea that physiological diversity is fascinating and interesting in and of itself. Curiosity regarding the varying physiological mechanisms across species should fuel disciplinary focus.

Krogh argued that a range of species should be studied from a physiological perspective for two reasons. First, one never knows when and which species will be useful for a scientific investigation. Second, species differences—in and of themselves—are interesting. Krogh's pronouncements in 1929 elide rather easily with the biodiversity discourse that has been developing since the 1980s. Conservationists frequently argue that the diversity existing on the planet today should be saved because we don't understand it, and it might one day be useful. But conservationists often add to this that it is also personally enriching to experience that diversity.[26] In this context, zoos achieve their scientific identity by understanding the physiologies of the endangered animals that they must know in order to save. The spectacle here is the diversity of life forms, which are appreciated through a greater understanding that is facilitated by technoscience.

It must be emphasized that basic scientists working in zoos continue to use domestic animals as "surrogate models" (Bolker 2009) of the endangered species in question. As already mentioned, zoo animals have long been difficult research subjects. In response, zoo scientists conduct initial research with standard animal models. Domestic cats are used as models of wild cats, dogs as models of wolves and bears, domestic ferrets as models of wild ferrets, and common frogs as models of wild amphibians. However, this initial research is followed by asking if and how domestic and wild species are similar and/or different. In other words, zoo researchers make explicit the comparative component that is central to modeling (Ankeny 2007). Budhan Pukazhenthi, who works at Conservation and Science, the Smithsonian Institute's research center, described the comparative aspects of animal modeling in his work as follows:

> So the way we've done it in our program is that any big question that we have an interest in answering is tackled with the domestic cat. And once we've thought that "well, we've got some answers," then we go see if

we can extrapolate into other species. Sometimes it works out that way; sometimes we just, it does not happen. That's where models do have limitations in how it does apply to reproduction itself.

Interview (April 11, 2006)

Animal models continue to be used in zoo research, but species generality and diversity is made into an explicit empirical question. In this context, zoo animals are protected research subjects, whose bodies contribute to an understanding of the diversity of physiological forms and biological processes that exist on the planet. That knowledge is the basis for developing interventions through which these species may be saved.

The Spectacle of Basic Science

Any kind of life science research involving endangered wildlife requires funding. As we saw in the previous chapters, zoos and biotechnology companies have engaged in cloning projects in part to create publicity. Cloning an endangered cow or an endangered cat gets reported through the popular press. And these reports find captive audiences, who may in turn become benefactors of this research by becoming a stock holder, a zoo member, or a zoo donor. However, this means that charismatic animals are required as the subjects of cloning experiments. For either zoos or biotechnology companies to get the publicity that they desire, they need to clone an animal that the public will be excited to see.

Bill Holt summarized this approach to funding reproductive science research in zoos as follows:

> I know some of the people doing the mammalian cloning research in the United States reasonably well. They are quite open in saying to me that they work on mammals because they are nice and furry and can generate funding. They have to rely on approaching funders for financial support and you can't get funding for amphibians. They are quite open about what people's attitudes are and what people like to give money to in order to save. It's good headlines, and good for funders. Even if it's not actually, in practical terms or real terms, going to make so much difference to the conservation of the animals. So I can see why they do that.
>
> *Interview (May 20, 2010)*

Hype is believed to generate research funding for zoos. Frogs are not considered good technoscientific subjects for generating this kind of capital.

Holt is concerned that research resulting from this kind of funding does not necessarily translate into the conservation of the species. He pursues basic science funding, which he believes could create an infrastructure for doing scientific research in zoos that is not rooted in the spectacle. In this context, Holt has sought to make this zoo science "boring." He articulated this vision for "boring" science to me as follows:

> I've always tried to tell people that what we actually need is a reliable program that's actually quite boring. They use artificial insemination in the cattle industry all the time. The whole dairy industry is based on it. And you don't see headlines saying "we have a new calf born today." . . . What we need is the equivalent of the black-footed ferret program, which is now almost at the boring stage. It works. It has led to reintroductions. They can use frozen semen and get pregnancies and the whole thing works really well. But that's the only example of where that's happened.
> *Interview (May 20, 2010)*

The idea here is that basic scientific funding that is not rooted in the spectacle may better enable reproductive scientists to normalize their work within conservation, by making it boring.

I asked Holt if he is trying to do cloning in a "boring" way. He chuckled and said yes. Holt's decision to clone frogs has allowed him to downplay the spectacle that is so often associated with cloning. The species he has chosen to work with was cloned over fifty years ago. Most people think that there is nothing spectacular about a cloned frog. Indeed, many people—including Holt and Lloyd—thought it would be easy to clone. Here we see how Holt has articulated cloning in a manner that rejects the hype so often associated with technology development. Situating his endangered animal research vis-à-vis John Gurdon's frog as opposed to Dolly the Sheep has allowed Holt to symbolize his cloning research as boring zoo science.

The Spectacle of Life

The spectacle in basic scientific approaches to the physiologies and developmental processes of endangered species lies not in what science and technology can do, but rather in the bodies of endangered species themselves. The diversity of life forms is a source of wonderment here, valued not only instrumentally through the animal model paradigm but also intrinsically.[27] In this sense, the life sciences are continuous with the long-standing aesthetics of environmentalism, wherein the sublime experience of that which is found is emphasized over that which is made. The life sciences offer a set of practices for understanding, and thus appreciating, the wonder of nature. Indeed, the basic scientists I spoke with commonly articulated their professional and personal identity vis-à-vis a curiosity in and fascination with biological diversity. They defined themselves as "animal people." And their love of animals was driven by a love of difference. Extinction posed a distinct kind of problem for these versions of "biophilia" (Wilson 1984).

For example, Barbara Durrant articulated her appreciation for nature as follows:

> I'm just fascinated with animals, fascinated with the diversity of animals. Wouldn't it be awful if all we had were cockroaches, mice, cattle, pigs, and horses? Won't that be terrible? I'm just so fascinated with nature and don't understand people who aren't. The diversity is just mind-boggling and there's a reason for all that diversity.
>
> *Interview (May 16, 2006)*

The idea that the world would, and possibly could, become a very boring place without a diversity of animals populating the planet is put forth here. Living in a world with only domesticated animals—mice, cattle, pigs, and horses—is considered a terrible prospect, especially if the only exception to such man-made species are pests like cockroaches. There is also a fear that such a world is on its way, with the linguistic shift from "wouldn't it be awful?" to "won't that be terrible?" And this future world of upcoming homogenization is considered not quite right. The diversity that exists has a reason; it serves a function.

Meanwhile, David Wildt alternatively expressed his appreciation for nature in this way:

I'm fascinated with species. I'm an animal person. I'm fascinated with understanding new mechanisms, especially those that allow a species to reproduce. I think scientists have this conventional wisdom about reproductive biology and it's too narrow. We're missing out on a lot. So our philosophy is that there's a lot of endangered species out there for a variety of reasons. If we don't study them who will?

   Interview (July 19, 2006)

Wildt locates the problem of extinction somewhat differently than Durrant, as lost scientific knowledge regarding the diverse mechanisms involved in biological reproduction. Extinction is a problem because it represents lost knowledge. But the predominant biological paradigm, which is premised upon the idea that biological forms and processes are conserved across species, is also viewed as a problem here. To contend with these twin problems, Wildt makes it his responsibility to know endangered species on a physiological level, to record their biological physiology as different from other, well-studied species. Basic scientific approaches are rooted in using and transforming the animal model paradigm so as to understand biological diversity, as a form of appreciation.

In social theorizing regarding modernity, science and technology have long been positioned as rationalizing and thereby dominating nature.[28] This idea has provided a basis for critiquing the loss of nature, most notably in Romantic thought and its twentieth-century reinterpretation through an ecological frame rooted in habitat destruction and species extinction. However, Nigel Clark (1997: 86) has noted that the rational and expressive elements of experiencing nature have long been entwined in science itself, most notably in the collection and classification practices of natural historians that undergird the modern zoological park. Basic scientific approaches to the reproduction of endangered animals are also embroiled in this history, where the expressive elements of experiencing nature occur in experimentation itself. Where technology development fosters public interest in nature through the spectacle of the clone, basic scientists experience the wonderment of nature themselves through the knowledge they produce as part of the life sciences.

Nature and Culture in Basic Science

It is interesting that one of the world's leading critics of cloning endangered animals is using nuclear transplantation at this historic moment, when the validity of cloning has come more and more into question. It is generally acknowledged that there are significant problems with somatic cell nuclear transfer in general and interspecies nuclear transfer in particular. As a consequence, its future use in either reproducing animals or developing regenerative medicine has been seriously called into question.[29] Nature has increasingly come to be seen as a barrier to somatic cell nuclear transfer. Biology is not as plastic and amenable to control as envisioned by Dolly's creator.

This seeming paradox is in part explained by the role of surprise and uncertainty in science itself. Hans-Jörg Rheinberger (2010) has, for example, argued that biotechnology is not simply an outcome of science. Rather, biotechnologies are an integral part of contemporary scientific knowledge practices. As the ability to clone endangered animals using interspecies nuclear transfer has been increasingly called into question on technical grounds, new kinds of biological questions have been raised. Disappointments with cloning have not brought cloning to an end. Rather, these disappointments have *contributed to* the sustained use of this technique in asking and answering new questions. Interspecies nuclear transfer has thus been rearticulated to again address questions about embryonic development within and across species. Cloning experimentation has fostered surprises, enabling new kinds of questions to be articulated about the complexity and diversity of life itself. The next chapter will explore in more depth these diversities, and the salience of appreciating the diversity of biological entities and processes involved in development for zoos.

But first, it needs to be emphasized that cloning does make nature in the zoo. One could interpret basic scientific approaches as embodying the idea that nature comes before culture because the found is prioritized over the made. I would, however, counter this interpretation by emphasizing that nature is rendered through the knowledge practices of basic life science research conducted in the zoo. Rheinberger (1997, 2000, 2010) has, for example, also emphasized that nature does not come ready-made to the laboratory. Rather, nature must be remade as research material, and is rendered meaningful through laboratory practices that occur within the histories of scientific disciplines. There is thus a path dependence to knowledge that is necessarily socially mediated. Biological facts cannot be separated from the milieu in which those facts come to make sense.[30] Nature is thus made in the zoo laboratory, but less as an object per se than as an object of knowledge. Interspecies nuclear transfer is a model system for making nature in this manner; technoscience is thus a way to appreciate nature as biology. And biology must always include both the thing being studied and the knowledge practices that render that thing meaningful (Franklin 2001).

7

Biodiversities

This chapter examines the varying logics of diversity that underpin basic scientific approaches to endangered animal reproduction in zoos. On the one hand, basic scientists want to understand the physiologies of wild animals on the basis that they are different from other, well-known species. In this context, extinction represents a potential loss of valuable knowledge regarding the diversity of life forms on the planet.[1] On the other hand, basic scientists also want to preserve that biodiversity. The goal is to use basic knowledge as a means to keep a range of different kinds of animal species on the planet. This twin desire to know and to save biological diversity was frequently articulated through a love of animals and of nature, which is a defining feature of "biodiversity" more generally.[2]

Here cloning is a technology for understanding, and thus appreciating, the diversity of life forms populating the planet. The technique could, for example, be used to compare the processes of embryonic development across different animal species; such comparisons are, indeed, a key feature of animal modeling itself (Ankeny 2007). But the technique is also, and more centrally to date, being used to understand the diverse entities involved in embryonic development. Holt and Lloyd's research is thus part of a growing attempt to understand the relationships between nuclear DNA and other biological elements within the cellular environment that shape developmental processes. This articulation is thus part of the displacement of nuclear DNA as deterministic in favor of more "relational" approaches to biology, which has marked much of the post-Human Genome Project era.[3]

This chapter situates the use of interspecies nuclear transfer within these relational approaches to biological developments.[4] I ask what this research agenda means for the ways in which zoos currently mediate the meaning of nature itself. Specifically, the genetic values pursued through selective breeding represent a different approach to genes and genomics when compared to those seen in emerging biological sciences, which are rooted in relationalities and complexities. Where genetic value locates "endangered species" in shared genealogy, more relational approaches to biology are increasingly questioning the role of shared environments in the processes of genetic expression. Where the focus on shared genealogy allows both zoo and wild environments to be more or less marginalized in determining what and who counts as an endangered animal, the focus on extragenetic factors in development works to highlight the spaces in which development occurs. In other words, the environment is understood here as constitutive. Nature does not have a genetic foundation, but is instead constantly in process, through relations that occur at varying levels. What does this mean for the genetic values of zoos?

Beyond the Genetic Determination of Difference

The trajectory of interspecies nuclear transfer has been bound up in the shift toward more relational models of biological development. In Holt and Lloyd's research, cloning is somewhat ironically being articulated in a manner that rejects the notion that genes are independent and deterministic. Other elements within the cell are viewed as importantly involved in reproductive processes. In this context, interspecies nuclear transfer is not only a system for making genetic diversity. It is also a model system for understanding the diversity and complexity of life itself.

Holt and Lloyd's research question is thus bound up in contemporary shifts within molecular biology. At the turn of the twenty-first century, it was generally assumed that protein coding DNA was the only kind of DNA that mattered for development. On this basis, grand claims were made wherein genetic sequences were equated with the "book of life."[5] Indeed, the notion of cloning as copying has relied upon this kind of gene fetish. And it was on this basis that mitochondrial DNA was deemed insignificant for development (Keller 1995). However, scientists are increasingly finding that noncoding RNA plays an important role in cellular development, which has destabilized genetic reductions.

In turn, the focus of molecular biology has been shifting from mapping the structure of genomes to asking how genes are switched on and off in the developmental mechanisms of cells and organs (Lock 2005: S47). As such, many geneticists no longer assume that genes determine phenotype in a straightforward, causal manner. Other elements, such as noncoding RNA and mitochondrial DNA, are being reopened for investigation. Epigenetics has become the new catchword in contemporary molecular biology.

This is not to say that nuclear DNA is not significant for biodiversity on the planet. Indeed, Evelyn Fox Keller (2010: 78–79) has emphasized that the turn to epigenetics and other more complex models of genetic expression does not diminish the fact that some changes in the base sequence of DNA along with some protein errors do have a significant effect on traits (e.g., single gene disorders). However, Keller contends that in the contemporary milieu it can no longer be assumed that the trigger of a trait is genetic. The environment—which can include other elements within the cell, the body, or be external to the body—can trigger DNA and produce genes in relation to other cellular acts. And one of the more spectacular findings in epigenetics is that these environmental factors that trigger DNA, but that do not change DNA, can be passed down to one or more generations.[6] Genes are here understood as expressed in relationship to other "environmental" processes, and these processes can be inherited.

While conducting the bulk of this research in 2005–2006, the people I spoke with rarely mentioned the term epigenetics. On the rare occasion when this word was used, it was generally to argue against the use of interspecies nuclear transfer on the basis that domestic animal DNA would be introduced into endangered animal bodies, which could have consequences for the classification of the animal in question, as seen in chapter 1. And Holt and Lloyd's research using interspecies nuclear transfer is not part of epigenetics. However, Holt and Lloyd's broader body of research does indicate an increasing engagement with epigenetics as a field of knowledge. For example, an article written by Stuart Milligan, Bill Holt, and Rhiannon Lloyd (2009) considers the potentially deleterious impacts of climate change on wildlife reproduction. The authors note that the field of epigenetics has consequences for conservationists who are concerned with these changes, in that environmental change could alter the reproductive viability and more basic characteristics of wild animal populations. Milligan, Holt, and Lloyd (2009: 3317) state:

> While the main drive to understanding "developmental programming" as described above came from its implications for human health, application of the concept to wildlife raises the possibility that any environmental (including nutritional, stress, pollutant exposure and disease) impact experienced by parents may induce epigenetic responses in the offspring. These could then contribute to the different characteristics, lifestyle pathways, and fitness of animals born.

Here we see the idea that genetics and environment combine in developmental processes, so that any kind of environmental change could introduce epigenetic alterations. In other words, changing environments change species in heritable ways. And these changes may affect whether or not species will be able to evolve.

What might the consequences of these changing knowledge practices be, given the role of "genetic values" within zoological parks? In many ways, it is simply too early to answer this question empirically. However, there are some clues that might indicate the kinds of opportunities epigenetics represents. Specifically, epigenetics may provide a new way for zoo scientists to link their concerns with genes and species to field conservationists' interests in ecosystems and habitats.

There is something of a divide between field- and lab-based scientists working in conservation. In his sociocultural analysis of the emergence of "biodiversity," David Takacs (1996) notes that "endangered species" was dislodged as an organizing principle amongst conservationists in the 1980s. With the shift to "biodiveristy," federal government agencies and environmental organizations increasingly supported efforts that operated at the level of the ecosystem or the habitat.[7] Survey techniques have been predominantly used to quantify the number of different kinds of flora and fauna within an ecosystem or habitat. This has meant that both zoos and lab-based researchers have been marginal to mainstream conservation efforts.[8] Indeed, the reproductive scientists I spoke with commonly commented that there is a chasm between biologists working at the level of the ecosystem in the field and life scientists working at the level of genes, individuals, and populations in the zoo. Most believed that those working at the level of the ecosystem continued to have greater power in defining the field.

Holt told me this story to exemplify this tension:

> We still have to cope with this level of suspicion around the place. Even yesterday, in fact, I have a student who is very keen on otters. She wants to understand the reproductive biology of otters and she's—that's her whole life. And she's shown that there are problems, people know little about the reproductive biology of otters, then there are collections where the otters don't breed and they don't know why not. So we put in an application to one of the conservation committees here, saying this is what we want to do. And they've actually said, okay, that's fine. But we don't consider this to be conservation. I really find that quite strange.
>
> *Interview (May 20, 2010)*

Studying the reproductive biology of otters is here considered a perfectly fine scientific project, one that the zoo approved. However, this research was not considered part of "conservation" by the zoo, and was thus also marginalized. I asked Holt why basic research regarding the reproductive physiology of an endangered otter does not count as mainstream conservation. He replied:

> I do see it as a culture of conservation biologists who are not lab-based. So there's a whole load of them who have no idea about what goes on in the lab. So if you tell them that something's happening in a lab they get very jumpy about it. Even if it's very simple, I think it scares them and I think that's what's behind it. They're much happier to do landscapes, to predict things statistically. [And] a lot of the big decisions in conservation are all owned by that community. I think that's why it's how it is.

Lab-based research is understood to be at odds with the statistical research methods of most conservation biology. In this context, it has been difficult to connect the research interests of those who work at the level of genes, individuals, and species with the priorities of those who work at the level of ecosystems, habitats, and regions.

With this tension in mind, it may be the case that epigenetics could provide a link that connects the laboratory with the field along with the molecular researcher to the ecosystem. Through the language of epigenetics, changes in the ecosystem could be read in the epigenome of endangered species. Epigenetics makes it possible to demonstrate how the environment gets into endangered animal bodies, something that has likely been suspected but has nonetheless been difficult to prove mechanistically.[9] This could prove particularly valuable in the case of endangered amphibians, where traditional survey methods have not been able to determine what is causing these species to decline. In lieu of known habitat problems, epigenetics could be used to ask if and how changing environmental conditions are changing the reproductive biologies of these species. Specifically, if the transpositions of technology developers seek to change endangered animals from the "inside out" (Clarke et al. 2003; McKibben [1989] 2003), basic scientists could transform endangered animals from the "outside in" (Landecker 2011: 178) through epigenetics.

Regeneration

Interspecies nuclear transfer initially developed in both human biomedicine and animal conservation through a desire to regenerate bodily parts and genomic bodies. The word regeneration means "reproducing parts of the organism which have been destroyed or removed" (*Oxford English Dictionary* in Franklin 2007b: 210). With humans, regenerative medicine holds the promise that one's cells could be used to regenerate one's organs in order to intervene in the processes of aging and death. Here the focus is on enhancing the individual body in an attempt to extend the individual life course. With endangered species, regenerative conservation holds the promise that the genomes of deceased endangered animals can be regenerated in order to intervene in the processes of population death, or extinction. Here the focus is on enhancing the population body, in an attempt to recuperate lost genetic information in order to stop a species from going extinct. Both humans and endangered animals are thereby bound up in the technical development of interspecies nuclear transfer and corresponding ideas about regeneration, which promises longevity that operates at the individual level with humans and at the population level with endangered species.

Somewhat ironically, there may be an "analogic return" (Franklin 2007a) to older definitions of "generation" as a consequence of "regenerating" endangered animals in the zoo. In pursuing the idea of inbreeding to its logical limit, somatic cell nuclear transfer has troubled the particulate notion of heredity upon which it is based. The environment—at least the cellular but also increasingly the wider environment outside the individual body—may again be perceived as playing a constitutive role in shaping the bodies of animals. This return will certainly not mean a wholesale reversal to prehereditary thought. Rather, the environment could become meaningful in new ways through the language of genetics and the "epistemic space" (Muller-Wille and Rheinberger 2007) of heredity. Specifically, the environment may be acted upon in a mechanistic manner to understand how genes are turned on and off, and with what physiological consequences. To elaborate upon this point, I will briefly discuss the notion of "generation" as it existed preheredity. I will then consider what an analogic return to generation, in the context of regenerating wild life, might look like, through the language of epigenetics.

As discussed in chapter 5, the notion of genetic value that currently predominates within zoos has been premised upon the idea that genealogy links animals that live in very different spaces. Producing genetic value relies upon many of the technologies associated with domestication, specifically selective breeding protocols that are often traced back to Bakewell's innovations during the eighteenth century. Genetic value is thereby bound up in the idea of heredity, one that was in formation during Bakewell's lifetime and that stablized in the middle to late nineteenth century with the work of Darwin and Galton.[10] Prior to the solidification of heredity, nature and culture were not viewed as separable or hierarchical in the processes of "generation." Staffan Müller-Wille and Hans-Jörg Rheinberger (2007: 3) have described the concept of generation as follows:

Until the end of the eighteenth century—according to some even until the advent of advanced cytological observations and Mendelism around 1900—hereditary transmission was not separated from the contingencies of conception, pregnancy, embryonic development, parturition, and lactation. Similarity between progenitors and their descendants arose simply because of the similarity in the constellation of causes involved in each action of generation.

Epigenetics is one site where the particulate notion of heredity, which gave rise to different forms of genetic reductionisms, is being problematized.[11] However, Margaret Lock (2005) has argued that epigenetics cannot be heralded as a wholesale rejection of genetic determinism. She develops this argument by considering the role of epigenetics in Alzheimer's disease research and care. Lock shows that epigenetics has certainly been taken up in this field, and has challenged the idea that a certain gene causes one to develop Alzheimer's disease. However, Lock also shows that much of epigenetics has continued to be located at the molecular level, focusing on how genes get turned on and off at the cellular level. In this context, environmental factors existing outside the individual body are rarely addressed. Meanwhile, Lock contends that multiple discourses can coexist in such a way that molecular biologists may reject genetic determinism while other scientists, clinicians, and public groups continue to hold onto deterministic repertoires. In this context, Lock warns that epigenetics in Alzheimer's disease research is a form of neoreductionism that forecloses questions about how human relationships shape the brain over time.

Meanwhile, Hannah Landecker (2010) has argued that the question of neoreductionism is yet to be determined in the case of nutritional epigenetics. Landecker (2010: 21) notes that nutritional epigenetics enacts a discursive shift from "food as fuel" to "food as information," wherein "bodies show at a macroscopic scale events happening at a molecular scale." For example, the macrolevel Dutch famine can be read in the epigenome of people who were in utero during that time. Landecker (2010: 25) notes that it remains uncertain as to how such scientific knowledge will inform governing regimes. It may be the case that food will be framed as personal choice through the language of epigenetics, in a manner that would reproduce maternal responsibility. This individualizing approach to the meaning of epigenetics would parallel the neoreductionism Lock finds in the epigenetics of Alzheimer's disease. However, Landecker has also pointed out that food could alternatively be framed as milieu, which would prompt greater social responsibility for "food environments" that people are located within but have little control over.[12]

Drawing on Lock and Landecker, I posit that the incorporation of epigenetics within conservation could link the lab and the field in a variety of ways. On the one hand, epigenetics could provide an extremely useful set of tools for examining the molecular consequences of macrolevel processes occurring in an ecosystem. Such technologies could help preservationists understand why certain species numbers are declining. This could be particularly useful in the context of amphibian conservation. However, epigenetics could also insert another kind of reductionism into habitat preservation practices, wherein molecularization displaces other knowledge practices based on the surveillance of different species that make up an ecosystem.[13] The genetic values of the zoological park would then become ever more hegemonic.

How might epigenetics shape notions of nature and culture in zoos and species preservation? David Takacs (1996: 102) points out that "ideas can cause ecological impact by reshaping how we view, and thus how we treat, nature." As such, by creating new ways of knowing and working on genes and the environment, epigenetics could also inscribe altered notions of what species are. On the one hand, such an approach could increase molecular definitions, expanding the terrain of genetics to cover not only species but also ecosystems. But on the other hand, epigenetics could alternatively inscribe an increasingly environmental definition of genomes and species. This would represent a possible fissure in the ways zoo animals are currently understood within zoological parks, as backups to endangered animals living in situ due to shared genealogy despite unshared space. The spaces in which animals of the same species live could again be understood as constitutive. Somewhat ironically, cloning could highlight the limitations of genetic bridges connecting endangered animals living in and ex situ.

Zoological Environments

Epigenetics provides a language for articulating the productivity of environmental factors for the phenotypes of animals. This is a language currently developing in molecular biology, denoting a shift within that scientific field. This language may provide new opportunities for life scientists working in zoos to link their research with mainstream conservationists' concerns. However, it must be emphasized that this linkage will likely draw upon already circulating ideas within zoos and amongst conservationists regarding the importance of in situ and ex situ environments. It is important to juxtapose the way the environment is being articulated through contemporary genetics with the ways in which zoos and conservationists alike have come to understand the productivity of environments for animals and, in turn, species.

Zoos have articulated their role in conservation efforts in many different ways. Zoos educate the public on the importance of conserving wildlife. They support field conservationists' work. They also support policy efforts that aim to protect habitats. But in addition, zoo animals have been considered useful for conservation efforts as backups and reserves to in situ populations. Zoo animals can be reintroduced into the wild as needed, providing a kind of safeguard for more precarious wild populations.[14] Like genetic value, reintroduction is based in heredity and assumes that genealogy connects animals living in different places. But is heredity a strong enough link? What other kinds of connections are required for zoo animals to be reintroduced into their native habitats?

Attempts at reintroduction have shown that captive animals are often ill-adapted at survival without their human companions. Keeping wild animals in captivity, providing their food and shelter, and controlling their reproduction are all domesticatory practices. While engaging with animals in this way may not change their genome, it does change the behavior of captive animals in a way that can have biological consequences.[15] For example, female cheetahs are solitary animals in situ, but are commonly caged in pairs within zoos. Close proximity with other female cheetahs appears to be associated with an increase in observable signs of social stress (e.g., pacing, growling, attacking, and fighting), which in turn leads to adrenal activation and increasing cortisol levels. These hormonal responses to stress have in turn been found to decrease ovarian function, which may explain why it has been so difficult to breed cheetahs in captivity (Wielebnowski et al. 2002). The environment of the zoo can therein delimit the biological reproduction of ex situ populations of endangered animals.

But captive environments have also been found to delimit the social reproduction of behaviors considered crucial to species themselves. In her classic monograph on monkeys, Thelma Rowell (1972) cites a study regarding the behavior of wild hamadryas (a kind of baboon) living in captivity. One of the distinguishing aspects of this species is that adult males herd females. Male animals born in situ and then moved into a captive setting will retain this behavior. However, males born in captivity will not continue this practice, which has been considered a species differentiating behavior by many (Rowell 1972: 80–81). Many people in zoos and conservation thus do not believe genes make a species; the environments that foster and/or delimit certain behaviors are also species-making.

As a consequence, reintroduction has continued to be an extremely difficult process for zoos and conservationists alike.[16] A metareview has, for example, found that reintroducing carnivores tends to be more successful when using wild-caught as opposed to captive-born animals (Jule, Leaver, and Lea 2008). And a study comparing reintroduction success rates among different generations of oldfield mice found that animals from a longer lineage of captive-born mice sought refuge less often than animals from a shorter lineage of captive-born mice. In other words, the longer an oldfield mouse's lineage had been born in captivity, the more likely the animal was to die from predation after being reintroduced (McPhee 2003). While zoos try to remediate their history by making rather than collecting their animal populations, zoo animals become reliant upon humans in a way that delimits their ability to live in noncaptive settings.[17]

Meanwhile, different modes of husbandry have also been linked with different levels of success with reintroduction. For example, Biggins et al. (1999) found that black-footed ferrets reared in outdoor pens were more successful in reintroduction when compared to cage-reared black-footed ferrets. The difference here was that pen-reared ferrets had to hunt prairie dogs that had established a burrow system with in the space of the pen, whereas cage-reared ferrets were either given meat or given a live prairie dog to kill. The ferrets in the outdoor pens were thus provided with an opportunity to learn how to hunt for food under monitored conditions, thereby improving their success in navigating their habitat upon reintroduction. Reintroduction thus requires creating suitable situations through which animals can transition from one environmental context to another, which is necessarily species-specific.

The problems associated with reintroduction, based on the limitations of a genealogical link, have had consequences for the frozen zoo. The genome of endangered animals may be regenerated through a combination of cryopreservation and reproductive technologies that include anything from artificial insemination to cloning. However, the frozen zoo does not provide a way to regenerate the behaviors of those animals, which to many people's mind is crucial in delineating what a species is and how it will survive into the future. The field conservationist I spoke with made this point as follows:

> [The idea behind the frozen zoo was,] "Look, don't worry about the wild. Things will be so hunky-dory in the frozen zoo. If they all go to hell in a hand-basket we can simply put them back in the wild." That's unrealistic because it didn't recognize that the behavior of animals was something you could not recapitulate through the genes. You had to conserve the behavioral integrity and even, if you like, the ecological integrity of that species, in order to have anything to put them back into.
>   *Interview (April 26, 2006)*

Here we see the idea that regenerating a genome is not sufficient because the resulting animal will not learn how to live in the world, as a wild animal. Any animal produced through the frozen zoo will require humans to teach it how to behave within the environment it was created for. Reproducing genomes is not the same as reproducing species.

Barbara Durrant drew on these discourses in order to argue that cloning endangered animals is problematic because it does not account for the behavioral aspects of species, which must be accounted for when deciding which endangered species are reasonable targets for such projects. She made this point by discussing why, appealing as it might be, cloning should not be used with giant pandas.

> There are more things to consider than just the technology of cloning. What do you do with those offspring? How would you raise them so that they would know they are pandas? How would you raise them so they

would have a chance of reproducing on their own? . . . We're not churn-ing out mice in a lab. We're churning out—we would like to churn out animals that can reproduce or even be reintroduced into the wild some day. So we have a lot of things to consider.

Interview (May 16, 2006)

Importantly, Durrant did not simply argue that interspecies nuclear transfer is technically underdeveloped. In addition, she argued that there was insufficient sociocultural knowledge available regarding how pandas live. This knowledge was considered crucial in order to ensure that regenerating the genome of pandas would recapitulate the ani-mals we know as pandas. Here Durrant extended the call from basic scientists to understand the reproductive physiologies of endangered animals. In addition, scientists need to understand the reproduc-tive practices of endangered animals. These practices are learned in a sociocultural milieu from other animals, and are not determined by the genome.[18]

I noted in chapter 1 that there was a missing position regarding the classification of clones. If articulated, that position would have said that cloned animals are not part of the endangered species—but for reasons other than their mitochondrial DNA. While no one articu-lated this position in relation to our conversations about chimeras, I think that the comments made by Durrant and the field conserva-tionist above give a good sense of how it might be done. The use of domesticatory practices in zoos, including not only reproductive tech-nologies but all the other technologies associated with captivity (e.g., housing and feeding), play a constitutive role in the social and biologi-cal lives of captive animals. These practices make zoo animals different from animals living in situ. Both the field conservationist and Durrant are thereby arguing that the environment in which endangered ani-mals are made matters at a material, sociological, and semiotic level. Mitochondrial DNA is a rather small mark of a far more significant set of relational practices occurring within zoos. Epigenetics puts these environmental influences, which are already being reckoned with in zoological parks and amongst conservationists, into a molecular and mechanistic language.

The Zoo as Nature

Emphasizing the environmental forces involved in shaping species could result in a reiteration of the long-standing critique of zoos: keeping "wild" animals in captivity is fundamentally wrong because the animals are no longer truly part of their species. This reinstates an essentialized version of wild animals as free and other to humans, which has long been used to argue against zoos and to position domestic animals as inferior (Haraway 2008).

I do not think that this is the only, or the most interesting, direction that can be taken when considering the constitutive role of environments. Rather than critique zoos, I think it would be more interesting to ask how cloning—and the zoo more broadly—might be rearticulated to incorporate the concerns of those who are interested in the constitutive role of the social and physical environments in which animals live.[19] This is, after all, the direction that Lloyd and Holt's cloning research initiates. In focusing on the role of mitochondrial DNA in early cellular development, they are using interspecies nuclear transfer to ask questions about the role of the cellular environment in development. They are also, elsewhere, posing questions about how larger environmental factors may have biological consequences that reshape reproductive processes. I interpret their work as extending the pluralist approach taken with the banteng cloning project, indicating how cloning specifically and the reproductive sciences more generally could be articulated with those who are most centrally concerned with the constitutive role that different kinds of environments have on animals in both social and biological terms.

Certainly some reproductive scientists working with zoos are already articulating their work in this manner. For example, Barbara Durrant described how reproductive research could be taken forward so as to include the physical and social environments in which animals live. She stated:

> It's our obligation to find better ways to get these animals to reproduce, more naturally. . . . Focus on giving the animals what they need to reproduce on their own. And that could be better enclosures, better nutrition, better medical care. Find out in your fieldwork how do these animals

live in the wild, what kind of sex ratios do they have in their groups, do the male or female offspring leave the group? All of these very basic things about the natural history of the species—incorporate them to the extent possible in a new situation or a captive situation and then do the minimum that is required. I mean ultimately if you have done an excellent job of providing the animals what they need socially, nutritionally, spatially they should reproduce.

> *Interview (May 16, 2006)*

Durrant emphasized the need to look at the specific social and physical environments in which animals live in situ. This knowledge should then be used to create better captive environments for zoo animals, which should compel the animals to breed on their own. She thus argues that reproduction needs to be understood more broadly to incorporate the physical environment to which animal behaviors and social organizations respond. The reproductive sciences are here articulated with the long-standing concerns of animal behaviorists, who can help translate the various in situ environments into captive settings.[20]

But there is also room for geneticists to ask questions about the relationships between zoological environments and reproductive processes. For example, Landecker (2011: 181) has noted that life scientists are increasingly studying the impact that food has on experimental animals, as inbreeding is not considered sufficient for creating similar kinds of animals. One could easily extend such questioning to ask whether and how the diets of zoo animals have epigenetic consequences.[21] Such research questions could be crucial if entire populations of endangered amphibians are being brought into captivity. Epigenetics could thus help conservationists understand not only how global warming is changing animals "out there, in the wild" but also how captive conditions change animals "in here, within the zoo."[22]

Such lines of questioning do not delegitimize the zoo as an unnatural environment, but rather highlight that the zoo is a particular kind of nature. Gregg Mitman (1996) has argued that the immersion exhibits developed in zoos enabled conservationists to see wildlife in new ways. The zoo functioned as a kind of laboratory for elaborating new conservation practices, wherein wild animals were surveyed at a distance. These visual practices coincided with the emergence of statistics in conservation, which emphasized counting animals in a habitat. The emergence of these practices was experienced as a loss by naturalists, in that it eclipsed the ideal of a chance encounter with wildlife in their habitat. As conservation displaced naturalism, and surveying life supplanted the randomness of an encounter, wildlife has increasing been transformed into an object of knowledge as opposed to a sublime experience. Nature was remade to be like the zoo.

In a play on Mitman's work, I can imagine Holt and Lloyd's articulation of cloning as ushering in a moment wherein "the zoo is nature." While the "unnaturalness" of the zoo has in a sense always been one of its defining features, this is now increasingly understood as continuous with rather distinct from the "unnatural histories" of most habitats and ecological niches that make up the planet.[23] For most conservationists, this represents an ontological shift and is experienced as a loss of other kinds of animals that can evolve in ways that are not mediated by humans. Here the zoo has stood for and symbolized the loss of a nature that is separate from humans. But an analogic return to generation, somewhat surprisingly, could alternatively enable the zoo to be understood as a kind of nature, as another environment in which different species live together.[24] In this context, the zoo does not have to symbolize loss but could instead offer a space for responsibly experimenting with better ways of living with other species within the constraints of specific kinds of natures.

Wild Life: The Surprises of Nature

What kind of wild life does the zoo as nature represent? Clark (1997, 1999) has described "wild life" as a mode of appreciating nature wherein engineered life is not made to be controlled but is instead meant to surprise its engineer. Where naturalists have long appreciated the surprise of a nature that is found, this imaginary instead appreciates the uncertainty of a nature that is made. As previously mentioned, the animals and organisms that people create often defy their expectations. Clark points out that these surprises have historically been feared. However, he contends that emerging imaginaries of nature are alternatively valuing the ways in which human-generated entities act back in surprising and world-making ways. According to Clark (1997), the film *Jurassic*

Park represents not only older fears about the risks of an unleashed nature but also points to our intrigue with such possibilities.

If the focus on transpositions amongst technology developers aligns with the salience of engineering in Clark's delineation of wild life, then the centrality of surprise in wild life aligns with basic science. Basic scientists appreciate the diversity of life forms as well as the diverse biological entities that shape the development and becoming of different animals. The goal here may not be to engineer animals anew. But this is nonetheless nature made through biotechnologies and other knowledge practices within the laboratory. Cloning has been articulated here to appreciate the surprises, specifically the uncertainty and the uncontrollability of biology as nature.[25]

However, the work of basic scientists also diverges significantly from Clark's articulation of wild life. Clark in many ways positions concerns about risks and hazards as passé. We shouldn't fear innovation, but should instead facilitate it. As a consequence, the surprise of biology is largely for innovation and novelty in and of themselves. This contrasts with basic scientists' very real concern that a number of species are endangered and are likely to go extinct. It also contrasts with the general belief within much of environmentalism that an unleashed technology is at least in part responsible for the current planetary state. As such, the appreciation of surprise is not bound up in innovation for innovation's sake with the frog cloning project. Rather, an appreciation for surprise and the unknown is instead bound up in questions about what constitutes responsible action. Following up on biological uncertainties may enable better ways of relating to endangered animals in different kinds of natures.[26]

CONCLUSION

We have produced the carbon dioxide—we are ending
nature. We have not ended rainfall or sunlight; in fact, rainfall and sunlight may become more important forces in our
lives. . . . But the meaning of the wind, the sun, the rain—of
nature—has already changed. Yes, the wind still blows—but
no longer from some other sphere, some inhuman place.
—Bill McKibben ([1989] 2003: 48–49)

The End of Nature is really After Nature: a point of appre-
hension, in this case of the constructional roles that that particular concept has played in our perceptions.
—Marilyn Strathern (1992a: 191)

[T]he certainty of what counts as nature—a source of insight
and promise of innocence—is undermined, probably fatally.
The transcendent authorization of interpretation is lost, and
with it the ontology grounding "Western" epistemology. But
the alternative is not cynicism or faithlessness, that is, some
version of abstract existence, like the accounts of technological determinism destroying "man" by the "machine" or
"meaningful political action" by "text." Who cyborgs will be
is a radical question; the answers are a matter of survival.
—Donna Haraway (1991: 153)

This chapter compares how the three different kinds of cloning projects described across this book seek to make nature in the zoo. How do
the different visions of nature, embodied by different cloned animals,

compare and contrast? What does cloning varyingly reproduce across different projects? What does cloning transform? How do different cloned endangered animals challenge us to think about and do "nature" in new ways? How do older definitions of nature get incorporated and remediated within these new iterations? How are such decisions made?

I use the three statements above to weave my way through these questions. It is generally accepted that the idea of nature as separate from human culture is no longer tenable, from either an ontological or epistemological standpoint. Where McKibben contends that this end of nature represents the finality of a particular kind of material planet, Strathern counters that the "end of nature" is actually the end of a highly constitutive idea. In this context, the question has become how to navigate a world wherein "nature" and "biology" can no longer be turned to for certainty in determining a course of action. This does not mean that technologies replace humanity, or animality. Rather, as Haraway states, the hybridities of nature and culture can be understood and acted upon in different ways; the question is thus how nature and culture are to be entwined in making the material-semiotics of the planet. How this question is addressed is nothing less than a matter of survival, a point made painfully clear by rampant rates of species extinction.

Drawing on these diverse areas of scholarship, this book has asked how cloned endangered animals embody particular ways of linking nature and culture. The focus has not been on separating out, or "purifying" (Latour 1993), the biological from the social parts of cloned animals. I have instead asked how the biological and the social, or the natural and the cultural, are varyingly understood and related in different cloning projects.[1] In the transpositions of technology developers, for example, culture is used to remake a failing nature. The end of nature here represents the end of biological limits, which has made it possible to make an improved nature in light of social and cultural conditions. In contrast, the genetic values of pluralist technology developers do not seek to change the genomes of endangered animals but instead use molecular definitions of species as guides to their social practices. The focus here is less on biological control as a mode of survival, and more on shepherding the genomes of endangered populations so that they may persist into the future. In this context, the genetics of endangered species are "strategically naturalized" (Thompson 2001) so that zoos can become a more ethical institution by remaking their own populations of animals. As such, the genetic values of zoological parks can be seen as aligning with the middle ground position that Paul Wapner (2010) has argued the environmental movement should pursue, wherein the idea of a pure nature and a pure culture are forgone in favor of pragmatic action. Finally, while the focus on biodiversities in basic science builds upon this shepherding ethos, the focus is less upon remaking populations per se and more upon facilitating the surprise of nature in and through science itself. Nature is here remade as research material in the laboratory so that different kinds of biological differences can be understood and appreciated. Following up on and facilitating surprises becomes the basis for forging responsible action here. As such, the three different kinds of cloning projects link biology and society, nature and culture in different ways. These links mean that the biological and the social "loop" (Hacking 1999) differently in the developmental processes of endangered animals. Nature is thus potentialized in heterogeneous ways by variously pursuing the potential of interspecies nuclear transfer.

Albeit in different ways, cloned endangered animals each embody the ways in which the nature of nature preservation has become explicitly "merographic" (Strathern 1992a). Nature is not understood or acted upon as separate from social life and culture in any of these projects, but is instead explicitly viewed as part of culture and part of something else, or part of nature. How nature is to be made in the zoo becomes the key question.

As such, cloning is not contentious in and of itself. Rather, the ways cloning pieces nature and culture together is the primary site of debate. Cloned endangered animals thus embody different "cultures of nature" (Macnaghten and Urry 1998). As Phil Macnaghten and John Urry (1998: 23) have pointed out: "There is no nature simply waiting to be conserved, but, rather all forms of its conservation entail judgments as to what indeed is nature." Different endangered animal cloning projects delineate the nature of nature preservation differently. The question of what world-making, scientific practices zoos should engage in is addressed in and through cloning itself.

Across this book, I have used "wild life" as opposed to wildlife to mark out the ways in which cloned endangered animals are both made by humans and yet are different from humans. Each cloning project picks up on different threads of Clark's (1997, 1999) delineation of this word. At first glance, the transpositions of technology developers appear most closely related. Clark focuses on science fiction and futuristic scenarios wherein humans make and engineer all sorts of new life forms. And the transpositions of technology developers are similarly rooted in this fascination with engineering new kinds of animals that could not be born outside the laboratory. But Clark's delineation of wild life also diverges from the work of the technology developers in important ways. Clark highlights that biology is made to be productive, to do something surprising with its life. But the life produced through technology development has all too often been made to represent the ways particular labs can control biology. The life cloned animals live has not been rendered terribly productive as a consequence. In this context, pluralist technology developers have instead emphasized the productive aspects of the clone's life for zoo animal populations and the zoo more generally. They have been concerned with the things that human-created animals can do. Specifically, they are interested in how the cloned banteng can be compelled to produce genetic diversity. While this picks up on the productive aspects of wild life, it does not, however, necessarily embody the element of surprise that is so central to Clark's definition of the word. Interestingly, it is basic scientists who have emphasized the ways in which life remade in the lab can be surprising. But, importantly, innovation is not pursued for novelty's sake alone here. Surprise, a long-standing element in naturalists' ethos, instead becomes a way to pursue responsible action by asking new kinds of questions that could enable not so much new biological forms but rather better ways of living with different species in a nature that is everywhere.

I now consider what this reworking of nature and culture means for the ways we think about cloning and clones as: 1) sites of reproduction; 2) embodiments of the politics of science; and 3) "figurations" (Haraway 2008) in the changing landscape of biopolitics.

Cloning and Clones

While conducting this research a field conservationist asked what exactly interested me about endangered animal cloning projects. "Do you want to know what is going on right now, on the ground, with this effort?" If so, he thought that cloning efforts were rather sporadic and the usefulness of this technique extremely uncertain. "Or, do you want to know what cloning might mean for the future?" If so, his vision was that cloning could radically transform our notion of what nature is and how we should go about preserving it. I responded that I was interested in both. I wanted to know how cloning was being used in the present in order to carve out new kinds of futures. The field conservationist paused. In that case, he responded, he would start by telling me about his own experiences with people who were cloning endangered animals. He would then tell me about what he thought cloning might mean for the future of conservation. And he would leave the relationship between the present uses of cloning and the possible future meanings of this technology up to me.

Like many other biotechnologies, there is a fissure in the ways cloning is discussed vis-à-vis the present and future. On the one hand, it is generally acknowledged that there are significant problems with somatic cell nuclear transfer and so its future use in reproducing animals, developing regenerative medicine, or conducting basic science has been called into question. On the basis of all these technical problems, human cloning has been outlawed in countries like the United Kingdom (Jackson 2001, 2009). On the other hand, there is still an ongoing sense that cloning is an inevitable part of our future. The concern here is that technical problems associated with cloning will eventually be worked out. And at that future point in time there will be nothing to stop cloning from moving to humans. This is generally considered a biopolitical nightmare, for human cloning is thought to lead to the creation of a secondary class of humans available for commodification under totalized control.[2]

The fissure between present and future also shaped my discussion with this field conservationist. Did I want to know about his involvement in very uncertain and tentative cloning projects occurring in the present, which did not seem to have very much potential? Or did I want to speculate with him about the ways in which cloning and other biotechnologies had the potential to radically revise the ways we think about nature and its preservation in the future?

Across this book I have drawn upon a nascent anthropology of potentiality to consider the presents and futures of cloning. This scholarship extends the sociology of expectations, which has argued that future promises regarding biotechnologies are constitutive of the present.[3] Specifically, promissory futures galvanize varying social actors and generate necessary capital investments. What this body of scholarship has shown is that we can use empirical methods to explore the kinds of worlds being carved out for the future with biotechnologies today. There does not need to be a fissure between the present and future in our discussions of what cloning and other biotechnologies mean. But where the sociology of expectations has highlighted the social and financial aspects of future orientations in bioscience and biomedicine, the anthropology of potentiality focuses instead on cultural and material processes (Svendsen 2011). Here the plasticity of biology is understood and acted upon through particular cultural tropes, which in turn works to delimit what biology can be made to do.[4] This analytic has been used across the book to argue that the cultural tropes used to delineate the potential of cloning (e.g., transpositions, genetic values, and biodiversities) shape the material development of cloned animals and endangered species on both a material and semiotic level. Cloning and nature are thus potentialized in tandem, and in varying ways. Both society and the material planet are made in and through cloning itself; they are "co-produced" (Jasanoff 2004; Reardon 2001).

What this highlights is that cloning is a rather flexible technique. It can be used in a number of different ways, and therein takes on a variety of different meanings. This is not terribly surprising from the perspective of Science and Technology Studies, but it is nonetheless important for the ways in which the ethics of cloning are discussed. If we want to know what cloning means and what kinds of ethical quandaries the technique poses, we need to consider the specific contexts in which it is being used. Cloning and cloned animals can defy our expectations and surprise us.

Cloning as a Site of Reproduction

Different cloning projects stand for different ways of reproducing endangered animals and, therein, reproducing nature itself. Emily Martin (forthcoming) has noted that, since the word came into use in the seventeenth century, reproduction has meant "carrying the act of production forward again in time." She has stated that the resulting person stands as a witness to the social and biological relations that brought the individual into existence. Drawing upon this definition of reproduction, this book has asked how endangered animals are being carried forward in time through the production of clones. These cloned animals stand as witnesses to the different biological and social relations that brought them into existence.

To carry species forward into the future, the processes involved in the biological reproduction of (some) endangered animals are being transformed. These biological transformations occur within social spaces that include a range of human and nonhuman actors who are brought together in and through the cloning process. Different cultural tropes shape the ways in which these gatherings are understood and acted upon. The clone witnesses these biosocial relations.

Technology development has, for example, worked to emphasize speed in innovation. This discourse has resulted in particular kinds of cells being used in experimentation, specifically those that are both available and most likely to work. The discourse of technology development has thus shaped the kinds of nonhumans (e.g., gaur, African wildcat, and sand cat cells) gathered together in cloning experiments. It has also shaped the development of the resulting cloned animals, which stand as witnesses to scientists' ability to transform reproductive processes by transposing different species bodies and corresponding relations. Interspecies nuclear transfer, as a technology, is potentialized in this set of practices. The cloned animal stands as a witness to the potential of this technique to reproduce endangered animals in new ways, and the potential of specific groups of scientists to use these technologies as needed. This has enabled others to envision a future wherein humans are able to remake wild life—from the inside out—to withstand a human-dominated planet.

The ways in which cultural tropes shape cloning experimentation becomes clear when technology development is contrasted with the different discourses that have been used to alternatively shape cloning research. The banteng cloning project, for example, combined the discourse of technology development with that of genetic value. This discursive combination worked to emphasize that the speed of innovation needs to be tempered so that the logic behind technological innovation can be simultaneously worked out. In other words, it was not enough to articulate in language the potential of a technique for the future; the potential must also be articulated in practice and therein enacted in the present. This resulted in a different kind of cell to be used in the cloning experiment, specifically cells that came from a genetically valuable, male animal. As such, the discourse of genetic value has shaped the development of the cloned animal and (it is hoped) the endangered captive population. Here the cloned animal is potentialized, as he could create genetic diversity in the captive population through his own sexual reproduction. The cloned banteng thus stands as witness to both his genetic heritage in relation to the captive banteng population alongside the potential of the cloning technique to reproduce captive populations in new ways. This represents a different imagined future, wherein animals are not remade from the inside out but are instead shepherded by humans in navigating a changing planet.

The discourse of genetic management has also been combined with that of basic scientific research, as seen in the amphibian cloning project. This discursive combination is interlinked with the ways cloning has been used to understand normal reproductive processes by introducing biological changes. The goal here has not been to reproduce an endangered animal per se, but rather to understand the diversity of biological processes involved in the reproduction of a wide range of different species. Cloning thus represents a model system as opposed to a technical feat. Reproduction, as both a biosocial process *and* an expert knowledge regime, is potentialized here, as nature comes to be understood as an object of knowledge situated in historical, social, and cultural time. Surprise is the key discourse that shapes this potential. Understanding reproduction and reproducing nature are here entwined in extending the goal of shepherding wild life, wherein the environment is potentialized. This opens up a space for questioning the ways in which wild life are remade not from the inside out, but rather from the outside in. The goal here is not to introduce biological changes to endangered animals through the apparatus of technological control. It is instead to understand how biological changes occur through the unintended pressures of a particular environment, which can include a number of different environments, namely, the cellular, the bodily, the zoological, and the in situ environments.

Importantly, none of the cloned animals stand as a solitary figure as a consequence of their mode of reproduction. As a copy, it is generally assumed that the clone lacks kinship relations. Replicating one individual presumably does not create new relationships with other individuals. In this context, Emily Martin (forthcoming) has imagined that the human clone would witness isolated individualism and a diminished sense of the salience regarding social relationships. However, cloned endangered animals are in many ways witnesses to multiple—and some may even say excessive—biological and social relations. The clone has biological links with three different individuals across two different species: the somatic cell donor, the egg cell donor, and the surrogate. The clone has social (and given recent rethinking on domestication, quite possibly biological) links with both the scientists who brought the animal into being alongside those who are responsible for ensuring the animal is cared for throughout life. This would include zookeepers as well as the other animals living in captivity with the clone. And the "genetically valuable" cloned animals are also embroiled within the genealogical relations of the captive endangered population, made evident through the kinship charts that zoos use to genetically manage their populations. The deficit in social relations amongst cloned endangered animals speaks more to a deficit in social relations amongst zoo animals more generally. Specifically, both clones and zoo animals raise questions about the nature of the relationship between animals living in and ex situ. As a consequence, cloning endangered animals does not enact an "analogic return" (Franklin 2007a) to resurrection. We instead see an analogical return to generation.

Rather than lacking biological and social relations, the biological and social relations of cloned endangered animals are variously acknowledged and acted upon. The clone's biological relation to a surrogate of a different species has been minimized as a consequence of the ways in which genetic inheritance has been understood as "biological" while pregnancy has been understood as "social." Similarly, labeling mitochondrial DNA as insignificant to the question of species relations has minimized the clone's biological relations to the egg donor of a different species. Minimizing the salience of these biological relations serves to minimize the salience of corresponding social relations, placing cloned animals in the social networks of endangered species. Highlighting these biological relations has alternatively worked to raise questions about the relationships between domestic animals, endangered animals, and humans.

Through the biological and social links with domestic animals, all the cloned endangered animals discussed in this book are embroiled in questions of enhancement and improvement. Martin (forthcoming) has argued that the hypothetical human clone would likely bear witness to a heightened desire for improving the stuff of nature. And this desire is certainly witnessed by each of the cloned endangered animals discussed in this book. Cloned endangered animals in turn embody a double anxiety that humans are "playing god." On the one hand, we see the fear that cloning could move from animal to human bodies, making humans into the "gods of their own creation." But environmentalists and some zoo workers are also concerned that humans are becoming the "gods of other species' creation." Both these anxieties are rooted in a worry that an increasing number of species are becoming the products of human manipulation, or domesticates. And domesticates are here generally viewed as inferior to wild animals.[5]

That said, while everyone I spoke with wanted to improve endangered species by making these populations larger and more genetically diverse both in and ex situ, the ways in which improvement was envisioned differed across projects. It is, for example, qualitatively different to improve the lot of endangered species by genetically engineering animals to live in a human-dominated world when compared to managing the genetic makeup of a population. Neither of these practices is rooted in a nature that comes before culture. Both are cultural activities that will contribute to the ongoing coevolution of the species in question. None of the animals could really be said to be wildlife. However, genetic engineering and population management engage with nature in very different ways. Where genetic engineering tries to change the biology of endangered animals, population management tries to keep biology more or less the same. Where genetic engineering sees biology as a problem to be overcome, population management uses biology as a guide. Where genetic engineering focuses on making biological changes occur from the inside out, basic science focuses on understanding how biological changes happen from the outside in. As a consequence of these differences, nature is made in different kinds of ways as an object of knowledge. And this has material consequences. On a very basic level, genetic engineering does not worry about the presence of mitochondrial DNA from a domestic animal in the genome of an endangered population. Mitochondrial change is thus allowed to happen. Population management does worry about this presence, and inhibits mitochondrial change from knowingly occurring. Mitochondrial change is not allowed to happen. There are thus different ways of reproducing endangered species and nature with cloning.

Clones as Embodiments of the Politics of Science

It is not uncommon in conservation—and in the sociology of the environment—to hear the refrain that debates over conservation are debates over what kind of world "we" want to live in.[6] Do we want to live in a world where humans remake endangered animals at the genetic level so that they better fit this human-dominated planet? Are we willing to become stewards in the genetic future of other species so that they may continue to live? Or are we willing to let species go extinct if they require such intensive levels of human care to make it on this planet? These are big and important questions, which emphasize both the thingness of nature and the ways in which our ideas about these things literally come to matter. As the field conservationist commented to me, debates about the world-making, scientific practices of zoological parks "are not questions for the technologists alone. These are much, much broader questions" (Interview, April 25, 2006).

But what kinds of forums are required to address such questions? Who is the "we" that gets to decide? Who is the "we" that wants a particular kind of world? One of the problems with this mode of questioning is that the "we" presented here is a unified, human "we." This "we" erases the fact that differently situated humans would probably like to live in different kinds of worlds. How to moderate such disputes? Additionally, this "we" also erases the question of what kind of world nonhuman animals might want to live in. How might we think about the interests and concerns of animals without resorting to anthropocentrism?

Across this book, I have countered the generality of the question of what kind of world "we" want to live in by instead asking what kinds of worlds are being actively made in and through the cloning of endangered animals. In the process, I have asked: who is actually involved in answering the question of what kind of world "we" want to live in, and who is excluded, through the act of cloning itself? Interestingly, the world-making activities that cloning endangered animals engages in have been disputed not only through public debates and commentaries in journals regarding whether or not endangered animals should be cloned. These disputes have also been conducted through cloning experimentation itself.

Cloning endangered animals is a site where disputes are actively engaged regarding the kind of world species preservationists are trying to reproduce. In delineating the notion of "matters of concern," Latour (2004b) has argued that it is not enough to criticize a technology; rather, one must engage with the technology so as to change it. He uses the example of environmentalists who scorn the SUV and refuse to drive it, arguing that they should instead become engaged with the SUV and the concerns of those who drive.[7] In many ways, the zoo-based scientists discussed within this book exemplify Latour's argument. While some critics of cloning endangered animals scorned the technique and refused to engage with it, others have articulated their concerns by using interspecies nuclear transfer itself. They have rearticulated cloning to address not only their worries about cloning but also the things that they care about regarding endangered species preservation. These scientists have thus engaged with cloning as a "matter of concern" (Latour 2004b) and as a way of expressing the matters that they care about (Puig de la Bellacasa 2011). They have engaged with cloning in order to articulate their vision of endangered species preservation, and the ways they would like to see different kinds of humans, nonhumans, techniques, and ideas brought together in order to make such futures possible.

In this context, cloning experiments have been a site where questions about science and politics have been addressed. Jenny Reardon (2007) has argued that the politics of science take on new meanings in the context of biopolitics. She contends that this is because the life sciences "make up people" (Hacking 2006), shaping the conditions of possibility for governing subjectivity (Rose 2007). Reardon (2007: 254) states:

> If we seek to build a science that better serves human ends, then it is
> not enough to understand how human concerns and interests shape

scientific ideas and practices. We must also understand how scientific ideas and practices help form a people with common concerns and interests. Democracy, as much as science, is made up, and the possibility for meaningful and reflective decisions in contemporary biopolitical worlds require that we consider and take responsibility for these intertwined, fundamental acts of world-making.

In the case of cloning endangered animals, these scientific practices form people and animals, creating modes of relating that are coconstitutive and are thus a matter of interspecies concern. Meaningful and reflective decisions regarding such world-making activities have occurred in the typical sites of public forums and debates along with opinion and white papers. But in addition, the politics of cloning have been engaged in and through experimentation itself. The concerns of some are here either actively engaged and acted upon or are actively dismissed. Democracy has thus been made up in the act of cloning itself. Technology development and pluralist technology development represent two different ideas about how the politics of science are to be delineated in the zoo.

Figuring Contemporary Biopolitics

Endangered species preservation is a site where nature has become synonymous with life, which has meant that nature itself has become understood as a biopolitical project. Zoo animals are made to live at a population level through selective breeding and at an individual level through improved enclosures, stimulation exercises, improved diet, and improved veterinary care.[8] Biopolitics is a grid of power relations that cuts across humans and animals. Biopower is part of domestication.

This book has shown how cloned endangered animals are embroiled with humans in the elaboration of regeneration, through which death— long the limit of biopower—is meant to be forestalled. Endangered animals have been cloned in part because biotechnology companies wanted to find out if they could use interspecies nuclear transfer as part of human embryonic stem cell research. This has created a link between endangered animals and humans in the elaboration of a biopolitical regime based on life extension, one that operates at the individual level amongst humans and at the population level amongst endangered species. Interspecies nuclear transfer marks in practice the cross-species development of "regeneration" through which both humans and nonhumans are being "optimized." Humans and animals are thus being remade, in tandem, through bioscience, biomedicine, and biotechnology. If social studies of bioscience and biomedicine are interested in the changing terrain of biopolitics today, the human cannot be separated out from the animal in our analyses.[9] Biotechnologies necessarily operate at the interface of human and nonhuman animal life. This was as true of the "older" biopolitics rooted in eugenics, wherein selective breeding was transferred from animals to humans, as it is of contemporary biopolitics, wherein humans and animals become together in and through the elaboration of biopolitical regimes rooted in regeneration.[10]

Questions regarding the biopolitics of cloning necessarily shift in this context. The concern is not necessarily that cloning will be used to mass-produce humans sometime in the future, creating a biopolitical nightmare in which surplus, second-class humans are made to enable the health of others. Rather, humans are being worked on in the very practices of cloning endangered animals. Biological regimes used to constitute one species are modified to constitute another. This does not mean that different species collapse into one another, becoming more or less the same. Indeed, the techniques and practices of facilitating life in agriculture, in the zoo, and in the clinic do differ. Rather, it highlights the ways in which bodies and techniques are being transposed to constitute multiple species in tandem, including humans.

The possible transfer of somatic cell nuclear transfer from animals to humans is not the key constitutive moment in cloning endangered animals. Rather, humans and animals are becoming together in and through the act of cloning. All sorts of traits may be unconsciously selected for in the process. There is an interspecies elaboration of a biopolitical regime rooted in regeneration, which seeks to reshape the lives of both humans and endangered animals. Humans are also engaging in domesticatory practices with endangered animals, which have long been socially and biologically constitutive for a large number of species. Humans are thus changing themselves in the process of cloning animals. This does not mean that humans should not clone endangered animals. Rather, humans should learn to respond well to the surprises that cloned animals create.

Notes to Introduction

1. Advanced Cell Technology is a biotechnology company that focuses on developing therapeutics for humans using stem cells as part of regenerative medicine. It was among the most famous and controversial companies in the debates over stem cell research in the United States during the Bush Administration.

2. Advanced Cell Technology created 692 embryos through this process, with 81 eventually developing into embryos (Lanza, Cibelli, Diaz et al. 2000). On this cloning project, see also Lanza, Dresser, and Damiani (2000).

3. See Fox News/Opinion Dynamics Poll (2002). This was a statistic that study participants commonly referred to during our discussions.

4. See also Haraway (2003a).

5. The future is a persistent point of reference in discussions about emerging developments in bioscience and biomedicine. This includes not only cloning and other reproductive technologies but also stem cells, neuroscience, genetic testing and modification, as well as biosecurity. Here, the future is represented as radically transformed by technologies that are imminent. This focus on the future has been critiqued as symptomatic of the excessive hype that so often surrounds these developments (Harrington, Rose, and Singh 2006). But scholarship has also made this 'future talk' an object of inquiry, as a key aspect of contemporary social life and social order. This literature has shown how imagined and desired futures are rendered knowable and realizable in the present (e.g., Rabinow 1999; Rose 2007; Brown 2003, 2009b; Brown, Rappert, and Webster 2000; Wyatt 2008; Sunder Rajan 2003, 2006; Adams, Murphy, and Clarke 2009; Geesink, Prainsack, and Franklin 2008; Van Lente 1993).

6. See Franklin (2003c) on how "life" is currently being recalibrated, an argument that I build upon in arguing that cloning endangered animals is a site where "nature" is being recalibrated as both an idea and as the materialities that make up the planet. See Clarke (1995, 1998, 2000) on just how controversial recalibrating reproduction has been since the beginning of the twentieth century.

7. There have also been attempts to use cloning to resurrect extinct species, such as the woolly mammoth. This endeavor had not been successful to date, and is not

discussed in this book. For a discussion of the hopes and fantasies surrounding the use of cloning to resurrect extinct species, see Turner (2002, 2008).

8. See Janssen, Edwards, Koster et al. (2003).

9. See Gomez, Jenkins, Giraldo et al. (2003) and Gomez, Pope, Giraldo et al. (2004).

10. See Gomez, Pope, Kutner et al. (2008). C. Earle Pope told me in an email (May 24, 2010) that the cat was being hand reared and died approximately sixty days after birth as a result of pneumonia. He questioned if this was the result of milk aspirated during bottle feeding. Another scientist I spoke with told me that there was an unconfirmed rumor that the somatic cell donor had a genetic disease. As such, how and why the cloned cat died is uncertain.

11. See Loi, Ptak, Barboni et al. (2001). For a discussion of this cloning experiment from an anthropological perspective, see Heatherington (2008).

12. See Chen, Wen, Zhang et al. (2002).

13. See Oh, Kim, Jang et al. (2008).

14. See Colen (1995), Davis-Floyd (1992), Franklin (1997b), Franklin and Ragone (1998), Ginsburg and Rapp (1995), Laslett and Brenner (1989), Martin (1987, forthcoming), Rapp (2000), Roberts (1997), Strathern (1992a), Stoler (1991), Svendsen (2011), Thompson (2005) and Yanagisako and Delaney (1995). However, it is important to emphasize that nothing—including, somewhat ironically, clones—is ever reproduced exactly the same (Strathern 1992a: 6).

15. See Becker (2000), Casper (1998), Franklin (1997b), Franklin and Roberts (2006), Mamo (2007), Rapp (2000), Roberts (1997), Strathern (1992a), and Thompson (2005). See Inhorn and Birenbaum-Carmeli (2008) for a review of this literature.

16. This reiterates, through technoscience, a long-standing relationship between wilderness and national identity. See Macnaghten and Urry (1998: 34–35).

17. See Franklin (2007b), Haraway (1989), Martin (forthcoming), and Ritvo (1987).

18. See Chen (2010), Ritvo (1987, 1997), Haraway (1989), and Thompson (2005). Each of these scholars points to the ways analogies between humans and animals are both gendering and racializing.

19. See also Thompson (1999), and Wilmot (2007).

20. See also Cassidy (2002), Derry (2003), Franklin (2007b), Ritvo (1995), and Thompson (1999, 2002a, 2002b).

21. See Callahan (1998) for a discussion of the ways ethical concern over cloning has been raised before and after Dolly the Sheep.

22. See the Wellcome Trust (1998), Williamson (1999), and Martin (forthcoming).

23. See Inhorn and Birenbaum-Carmeli (2008), and Thompson (2005) for a discussion.

24. On constructionist approaches to technologies, see Bijker, Hughes, and Pinch (1987), Clarke and Montini (1993), MacKenzie and Wajcman (1999), Saetnan, Oudshoorn, and Kirecjczyk (2000), Thompson [Cussins] (1996), Timmermans (2000), Winner (1980), and Wyatt (2008).

25. See also Saetnan, Oudshoorn, and Kirejczyk (2000).

26. See also Latour and Wiebel (2005).

27. Haraway is here drawing on Thompson's (2005) argument that assisted reproductive technologies are not really used to make babies per se, but instead to
"make parents." Here Thompson emphasizes the ways in which technologies are
ways of making relations.

28. On actor network theory, see Callon ([1986] 1999, 1987), Latour (1987, 1988,
2005) and Law (1999). See also Pickering's (1995) related notion of a "mangle."
On social worlds/arenas, see Clarke (1991, 1998, 2005), Clarke and Fujimura
(1992a), Clarke and Montini (1993), Clarke and Star (2007), and Fujimura
(1992, 1996).

29. I am here drawing on Latour's (2004b: 246) argument that science and technology studies detects *"how many participants* are gathered in a *thing* to make it
exist and to maintain its existence." (Emphasis in original.)

30. See also Franklin (2003).

31. This is the central argument of the sociology of expectations, which has
considered the role of rhetoric in the development of emerging technologies.
Future-oriented narratives commonly frame technologies like cloning, which
galvanizes financial and other kinds of support for their development in the
present. As such, the sociology of expectations has argued that prospecting
cannot simply be understood as hype, but is instead a constitutive force in the
social organization of the present (Brown 2003). For the role of the future in the
biosciences and biotechnology today, see also Brown (2009b), Brown, Rappert,
and Webster (2000), Rose (2007), Sunder Rajan (2006), and Wainwright, Williams, Michael et al. (2006).

32. I would like to thank Karen-Sue Taussig and Klause Hoeyer for organizing the
workshop on the Anthropology of Potentiality in Teresopolis, Brazil, in 2011,
which was generously supported by the Wenner-Gren Foundation. The papers
resulting from this workshop can be seen in a Special Issue of *Current Anthropology* (forthcoming). Many of the papers are similarly operating at the intersections of social studies of reproduction and science and technology studies.

33. See Robinson (1996: 32), Baratay and Hardouin-Fugier (2002), Rothfels (2002),
and Hoage, Roskell, and Mansour (1996).

34. See also Veltre (1996).

35. On contemporary changes in the display practices of zoological parks, see Baratay and Hardouin-Fugier (2002), Hanson (2002), and Rothfels (2002).

36. See Macnaughten and Urry (1998), and Wapner (2010) for a discussion.

37. McKibben ([1989] 2003) has been a vocal critique of biotechnologies in general.
For a discussion of his critique of genetic technologies for human use in biomedicine, see Franklin and Roberts (2006).

38. For a critique of this approach to nature preservation, along with an actor network theory based mode of breaking out of the trappings of nature/culture that
such arguments represent, see Latour (2004a).

39. On the end of nature in relationship to the materiality of the planet, see McKibben ([1989] 2003). On the end of nature as a concept for a unified whole, see Cronon (1997). For a synthesis of these two arguments that seeks a new way of thinking about and enacting nature preservation, see Wapner (2010).

40. See Western, Strum, and Wright (1994) for this argument within the context of conservation, Wapner (2010) for a reflection on the argument within the history of the environmental movement, and Thompson (2002b) for a sociological reflection.

41. For a discussion of the STS scholarship on nature and its consequences for biodiversity conservation, see Lorimer (2012).

42. See Haraway (1989, 1991, 1997), Latour (1993), and Shapin and Schaffer (1985).

43. See Franklin (2003a, 2003b, 2003c), Haraway (1989, 1991, 1997, 2008), Latour (1993, 2004a, 2004b), Rabinow (1996, 2000), Rheinberger (1997, 2000), and Strathern (1992a, 1992b).

44. See, for example, Haraway (1991), and Rheinberger (2000). To some extent, Strathern (1992a) also makes this argument. By being made explicit, Strathern argues that society, the individual, and nature are vanishing. What appears in these displacements is choice. Nature is no longer the grounds for family and society, but is instead something that is consumed. While the flora and fauna may have not disappeared, nature has nonetheless lost its analogical function. "If nature has not disappeared, then, *its grounding function* has. It no longer provides a model or analogy for the very idea of context. With the destabilising of relation, context and grounding, it is no surprise that the present crisis (epoch) appears an ecological one. We are challenged to imagine neither intrinsic forms nor self-regulating systems" (Strathern 1992a: 195). "After nature" delineates both the idea that culture comes after nature and its disappearance. But see Franklin, Lury, and Stacey (2000), and Franklin (2003b) for an argument that uses Strathern's notion of merography to argue that nature continues to be useful analytically, albeit in a different form.

45. Paul Wapner (2010) makes a similar argument, although through a very different theoretical tradition.

46. This is continuous with Macnaghten and Urry's (1998) definition of nature as social practice, which they contend allows different cultures of nature to emerge and thus be contested.

47. See also Clark (1997).

48. See Chrulew (2011) for an extension of biopolitical theorizing to zoo animals. See Lorimer and Driessen (in press) for a discussion of the biopolitics of different modes of human-animal relations, including rare breed preservation. See Shukin (2009) for a discussion of biopolitics in the context of animal capitalization, which she develops through the work of Hardt, Negri, and Agamben as opposed to Rabinow and Rose. It is important to emphasize that where Chrulew has been centrally focused on using biopolitics to critique what humans do to

animals in the zoo, I use this theory to ask how humans and animals become together in zoo science.

49. See also Thompson (1999).

50. For a discussion of how contemporary biopolitics and biomedicine are increasingly being focused on bodily and temporal transformation aimed at the future, see Clarke (1995), Clarke, Shim, Mamo et al. (2003), Franklin (2007), Franklin and Lock (2003), Landecker (2007), and Rose (2001, 2007).

51. On visibility and invisibility more generally in relation to articulations, see also Star (1991), Star and Strauss (1999), and Casper and Moore (2009).

52. My approach to studying cloning endangered animals is consistent with and extends the use of social worlds/arenas in studying scientific practices (Clarke 1991; Clarke and Star 2003, 2007; Clarke 2003, 2005). As Clarke and Montini (1993) have pointed out, this approach can accommodate the notion of nonhuman agency that is so well associated with actor network theory and science studies, which this book draws upon. However, I find interactionist approaches to scientific practice more analytically flexible than actor network theory. This "theory-method package" allowed me to consider questions of symbolism and meaning. It also advocates for considering what is left out of the relations that make up an articulation, which is crucial for studying power relations. On the limitations of actor network theory for feminist science studies, see Star (1991).

53. For an extended discussion of my use of these techniques, see Clarke and Friese (2007).

54. Franklin opens her book *Dolly Mixtures* (2007b: 1) by noting that we are just beginning to develop a suitable language for thinking about the meaning and consequences of cloning. Indeed, the people I spoke with often struggled to talk about cloning due to the lack of such a suitable language.

55. See also Franklin (2007b), Haraway (1991, 1997, 2008), and Thompson (2013).

Notes to Chapter 1

1. For an analysis of this discourse, see Franklin (1997a, 2007), Hartouni (1997), Petersen (2002), Nelkin and Lindee (1998), Nerlich, Clarke, and Dingwall (1999), and Wilkie and Graham (1998).

2. For a discussion of the human embryonic stem cell debates in the United States, see Benjamin (2013), Ganchoff (2004), Maienschein (2003), and Thompson (2005, 2013). For a discussion of the debates internationally, see Jasanoff (2005), Gottweis, Salter, and Waldby (2009), Liu (2008), and Prainsack (2006). For a discussion of how religion has shaped public discussion of reproductive and genetic technologies that includes cloning, see Evans (2010).

3. For a discussion of the public concerns over consuming cloned agricultural animals, see Gaskell (2000), and Priest (2000).

4. See Martin (forthcoming) for a discussion of this argument within the social studies of reproduction.

5. Historically, mitochondrial DNA has been considered insignificant (Keller 1995) and its role remains largely uncertain in genetic research.

6. Martin Johnson kindly pointed out to me that contemporary biologists would call the embryos and animals resulting from interspecies nuclear transfer "heteroplasmic." These cells and animals are "chimeras" according to traditional definitions of the word, and are referenced as such in the popular press and in my informal conversations. But these entities are not considered "chimeras" as the word is formally defined within biology today.

7. This chapter builds on a significant literature on classification across science and technology studies and cultural sociology. See Bowker and Star (1999), Epstein (2007, 2008), Foucault ([1966] 1970), Lakoff and Johnson (1980), Ritvo (1997), Shim (2005), Starr (1992), and Zerubavel (1991, 1996).

8. In this sense, interspecies nuclear transfer may be understood as a site where technologies are being used to engage in a variation of the horizontal gene transfers seen in other species. For a discussion of horizontal gene transfer, see Helmreich (2003).

9. The official classification of endangered species relies upon the assumption that classifications result from the physical properties of the object/subject in question, an assumption that has defined the contemporary episteme. One of the normative features of "modernity" has been the use of science and corresponding technologies to arbitrate and determine the truth of what something is and what it is not according to its physical features. This epistemological framework is often referred to as realism. Given the ways that realism pervades Euro-American social orders and its prominence in zoos, it is not surprising that almost everyone I spoke to while conducting this study sought to resolve the dilemma of what chimeras are through reference to the body of the cloned animal. Here, the goal was to use physical reality to determine social order. See Bowker and Star (1999), Douglas ([1966] 2005), Epstein (2007), Foucault ([1966] 1970), Shapin and Schaffer (1985), and Latour (1993).

10. There are long-standing critiques of realism that provide important analytical frameworks for understanding why mitochondrial DNA alone cannot resolve these debates. Social theorists from varying intellectual traditions have argued that understandings of the physical world are informed by cultural assumptions and social systems. This includes structuralist (e.g., Durkheim [1912] 1995; Durkheim and Mauss [1903] 1963; Douglas [1966] 2005; Canguilhem 1978), poststructuralist (e.g., Haraway 1989; Butler 1993), and interactionist (Mead 1970) critiques. Here we see the idea that classification systems do not reflect an unmediated physical reality, but rather collective ways of knowing. It is argued that these collective ways of knowing delimit and shape what can be known of the material world itself.

11. A number of scholars have pointed out that any classification requires emphasizing certain similarities and/or differences while marginalizing others. See

Bowker and Star (1999), Foucault ([1966] 1970), Lakoff and Johnson (1980), and Zerubavel (1991, 1996). In her ethnography of assisted reproductive technologies in the United States, Thompson (2001, 2005) showed that people would intertwine biological and cultural narratives in order to naturalize their kin relations when using assisted reproductive technologies, often in varying ways that had contradictory results. The biological was flexible enough to accommodate many different notions of family, depending on which biological processes were privileged and which were marginalized. I similarly found that the biology of chimeras was a resource for classifying chimeras. However, these biological properties were similarly flexible so that certain processes could be highlighted or hidden in schematic reference to hybrids and bridges. Chimeras were thereby rendered meaningful in multiple and contradictory ways.

12. See DiMaggio (1997).

13. See Brubaker, Loveman, and Stamatov (2004), and Zerubavel (1991, 1996, 1997).

14. See Bowker and Star (1999), Foucault ([1966] 1970), and Zerubavel (1991, 1996).

15. See Levine (2002).

16. See also Thompson (1999) on just how negatively defined hybrids are in zoos.

17. This logic has, however, been critiqued by conservationists. See Wapner (2010), and Western, Strum, and Wright (1994).

18. See also Thompson (2005), and Casper (1998).

19. See also Nelkin and Lindee (1995).

20. See Nelkin and Lindee (1998), Edwards (1999), Hartouni (1997), and Priest (2001).

21. Keller (1995) has provided a feminist critique of the longstanding tradition in genetics that has deemed nuclear DNA as solely relevant and mitochondrial DNA as insignificant.

22. Mitochondrial DNA is frequently used in studying human evolution (TallBear 2007), and is also used for human ancestry testing (Bolnick, Fullwiley, Duster et al. 2007; Nelson 2008).

23. It is important to point out that there was significant debate in the zoo field regarding the validity of this assumption across species. At the time of research, scientists were still assessing the mitochondrial DNA of clones and their offspring.

24. All other cloned endangered animals created in the context of zoo-based science have, to my knowledge, been male.

25. Seeing silence in the research situation is a core component of positional maps, as described by Clarke (2005). This is because these silences often point to sites of uneven power relations. See also Casper and Moore (2009), and Star and Strauss (1999).

26. Bruno Latour (1993: 41) has argued that undertaking hybridization requires a belief that serious consequences will not result. This belief can be created by bracketing off the social from the natural, or it can be facilitated by thoroughly thinking through the connections between the social and the natural so that

dangerous hybrids are not carelessly introduced. Drawing on Latour's work, I would say that this set of classificatory practices takes the latter approach.

Notes to Chapter 2

1. The scientists I spoke with at this lab took the position that interspecies nuclear transfer is a necessary route, and the problems associated with creating hetero-plasmic individuals were minimal compared to species extinction.

2. See Friese (2009).

3. The San Diego Zoo Global's research center CRES has trademarked the brand Frozen Zoo, which has since become a shorthand for these kinds of collections. When I refer to CRES's collection, I use the trademark symbol. When I refer to the more general practice, I do not.

4. The idea here was that human somatic cells could be transferred into cow eggs in order to make an embryo that was not "human," but from which "human" stem cells could nonetheless be derived. The company could thus work around religious opposition to human embryonic stem cell research, on the basis that "human" embryos were not being destroyed. In addition, the company would not have to collect human eggs from women's bodies. This allowed them to work around feminist opposition to human embryonic stem cell research, not to mention a serious technical challenge. See also Franklin (2003a) on these kinds of biotechni-cal work arounds to ethical dilemmas on the part of biotechnology companies.

5. Interview, Mike West (July 18, 2006); interview, Philip Damiani (July 1, 2005).

6. Nuclear transfer remains inefficient in terms of the material resources that are required to produce a limited number of embryos and offspring. The process requires both a large number of egg cells for the nuclear transfer procedure and a significant number of gestational surrogates for the development of resulting embryos. Despite these material investments, relatively few animals are actu-ally birthed. As seen in my description of the cloning experiments at ACRES, approximately 120 egg cells were used in one day of research. These retrievals were done two times per week for several years. Given the endangered status of endangered species, it is materially impossible and is generally thought unethi-cal to use endangered animal egg cells in the highly experimental processes of somatic cell nuclear transfer.

7. See Friese (2009).

8. See Hradecky, Stover, and Stott (1988).

9. Interview, C. Earle Pope (April 14, 2006).

10. It is not unusual to see diagrams of the cloning process that delineate the cells brought together to produce a cloned animal. This mode of representation is common in the life sciences, and presents the technical process of somatic cell nuclear transfer as a kind of "recipe" (Lynch and Jordan 2000). Good biologists have long been seen as having "golden hands," which are able to make these recipes work in practice. Building upon this language, the scientist from ACRES emphasized in our discussions that this laboratory did not simply follow a

cloning recipe. Rather, the recipe needed to be built upon and tweaked so that it would work in their hands.

11. Litters of male cloned wildcats were born in August and November 2003. Litters of female cloned wildcats were born in April and May 2004. See Gomez, Pope, Giraldo et al. (2004).

12. Betsy Dresser left ACRES in 2011 (Eggler 2011), but C. Earle Pope and Martha Gomez continue to work there as of 2012.

13. See Anderson (2001).

14. Adele Clarke (1987) has shown that the availability of research materials was a primary factor in shaping the direction and pace of the reproductive sciences. In the case of cloning endangered animals as part of technology development, the availability, familiarity, and likelihood of producing a living offspring were similarly the primary criteria for determining which animal should be cloned.

15. On the uneven politics of "mutual" benefits projects, see Hayden (2003).

16. I was conducting this research during and after Hurricane Katrina. Scientists at ACRES told me that they wouldn't be able to rely upon state funding and private donations to support their research to the same extent post-Katrina. In this context, these scientists were turning to medical funding.

17. Captive breeding is a key feature of domestication itself and many species resist the kind of control that this process entails (Franklin 2007b: 88). Getting zoo animals to breed in captivity, and of their own accord, is a serious challenge. As such, many zoos repeatedly bred zoo animals who were willing, a problematic practice that will be further discussed in chapters 4 and 5.

18. See Hradecky, Stover, and Stott (1988), and Stover, Evans, and Dolensek (1981).

19. See Hochadel (2011).

20. See Berger (2008) for a critique of seeing animals in zoos, which he contends is entirely lacking in spectacle.

21. See Hanson (2002), and Mitman (1999).

22. See Mitman (1999).

23. I would like to thank Stefan Timmermans for this point. He and his son looked the banteng up on YouTube one night during my postdoctoral fellowship. Timmermans reported back that his son's response was that the banteng was "just a cow."

24. See Brown (2003, 2009b), Franklin (2007b), Martin, Brown, and Turner (2008), Rose (2007), and Sunder Rajan (2006).

25. Wealthy benefactors have provided initial funding for companion animal cloning projects. See Klotzko (2001: 171) for a discussion of this in the context of the dog cloning, Missyplicity project.

26. The publicly available financial statement for the Audubon Nature Institute supports this, which states that the Audubon Nature Institute and Audubon Nature Institute Foundation raised $295,121 for the year ending December 31, 2007 in grants; meanwhile, they raised $4,607,840 in gifts and exhibit sponsorship. The report also states that ACRES, and the survival center within which it is located,

raised $1,411,201 in 2007 and $2,227,444 in 2006. While I cannot say how much money was raised through cloning endangered animals, it was clear that this and other organizations use spectacles outside of the park in order to garner private donations.

Notes to Chapter 3

1. See Hill and Dobrinski (2006).
2. See also Franklin (2006).
3. See Fuglie, Narrod, and Neumeyer (2000).
4. See Ankeny (2007), Bolker (2009), Logan (1999, 2005), Löwy (1992), and Rheinberger (2010).
5. See Rader (2004) on the commercialization of experimental mice in the history of the life sciences and biomedicine.
6. See also Logan (2002, 2005).
7. See also Ritvo (1996).
8. Domestic animals are easier to work with than endangered animals, but it would be a mistake to assume that domestic animals are "easy" to work with. While in the domestic cat colony, I witnessed one new cat resist testing procedures. When gathered from her room occupied with other cats, this cat broke free from the attendant's arms and began running back and forth along the corridor, all the while screeching. For what felt like minutes but was probably only seconds, the animal attendant, the lab tech, and I stood there taken aback. As the cat ran toward us, the animal attendant tried to capture her. However, the cat was able to break free by swiping the attendant's face. The attendant was eventually able to console the cat by using her voice, and was able to get the cat to stop running. Hovering in the corner of the corridor, the attendant wrapped a towel around the cat, and returned her to one of the rooms in the colony. The cat successfully avoided having her fertility tested, and thereby was not a candidate for egg retrieval and surrogacy the following week.
9. See Logan (1999, 2001, 2002, 2005).
10. My argument here builds directly upon Thompson's (1996) analysis of the ways in which patients in IVF have to move between subject and object positions for the technique to work, and for infertile patients to become parents. Biomedicine operates on the assumption that the body and its parts can be understood and acted upon as objects. Thus, physicians must "objectify" the bodies of their patients at certain moments in the biomedical encounter. This is not to say that physicians think of their patients as objects, but rather they must engage with patients as objects in certain instances. This is, however, moderated by the need to also subjectify patients. Thompson's important point here is that being objectified is not necessarily a problem. Some of the women she interviewed were willing to be objectified in order to become parents. Rather, problems arose when patients were inappropriately objectified within the ontological choreography that makes up the medical encounter. The ways in which patients moved

between subject and object positions are therefore crucial, and were thus highly staged and carefully organized practices. Drawing on Thompson's argument, I contend that the fact of ontological choreography itself is not a problem with cloning endangered animals. Rather, the ways different ontological positions are organized across the developmental process has at times been problematic. Indeed, it is not surprising that there would be problems with the ontological choreography of the first cloned endangered animal, which was necessarily a new and creative venture. Highly staged and carefully organized ways of managing the ontological choreography of cloned endangered animals had not yet been produced. Developing such an ontological choreography has been part of the work of cloning, albeit an underappreciated component.

11. See Lanza, Cibelli, Diaz et al. (2000).
12. According to C. Earle Pope, the cloned sand cat also died because of husbandry difficulties (email, May 24, 2010).
13. Sarah Franklin is here drawing on arguments made by Strathern (1992a), Rabinow (1996), and Haraway (1997), in which it is argued that nature and biology are no longer guaranteed a foundational position in ontological orders with the increasing culturing of nature and biology in the life sciences and biomedicine.
14. Cloning is itself this kind of mixture, previously thought to be a biological impossibility. See Franklin (1999, 2007b).
15. See Thacker (2003: 76) for a similar critique of the tool metaphor.
16. I am here referencing the book *The Right Tools for the Job* (1992a), edited by Clarke and Fujimura.
17. On constructionist approaches to technologies that emphasize power relations, see Bijker, Hughes, and Pinch (1987), Clarke and Montini (1993), MacKenzie and Wajcman (1999), Saetnan, Oudshoorn, and Kirecjczyk (2000), Thompson [Cussins] (1996), Timmermans (2000), and Winner (1980).
18. See Franklin (1997a, 2001, 2003, 2006), Franklin, Lury, and Stacey (2000), Haraway (1997), Latour (1993), Rabinow (1996), and Rheinberger (2000).
19. On this romantic notion of wildlife, see also Mitman (1996), Oelschlaeger (1991), and Takacs (1996).
20. See Western (2007), and Western, Strum, and Wright (1994).
21. These ideas have recently been made available for more popular consumption with the documentary *Unnatural Histories* (Nightingale and Murray 2011). See also Wapner (2010).
22. See also Clark (1997) for a theoretical discussion of the ways in which "risks" are reinterpreted in such projects.
23. See Takacs (1996) for a discussion of restoration ecology, and debates about these developments amongst conservationists.
24. Fuller (2011) has referred to the kind of human envisioned through transhumanism as humanity 2.0. I use endangered species 2.0 to emphasize continuities between the human-technology and animal-technology interfaces. But where Fuller uses his sociological research to implicitly support transhumanist goals,

I draw on feminist and other areas of critical social theory, including animal studies, to engage with this imaginary as only one way of making nature into the future. In the process, I highlight the power relations and political struggles that are effaced in both transhumanism as well as Fuller's discussion of this movement.

25. See, for example, Savulescu (2010), Bostrom (2004), and Savulescu and Bostrom (2009). For a discussion of transhumanism from a critical sociological perspective, see Franklin (2010), Newman (2010), Thacker (2003), and Twine (2010).

26. Twine (2010: 182) notes that transhumanists tend to see the use of technology as an ethical imperative.

27. Transhumanists tend to take as their counterargument those who see biotechnologies as eroding an a priori, unified, and fundamental human essence. For this argument, see Fukuyama (2002) in relation to bioethics, and Habermas (2003) in relation to sociology. As a consequence, transhumanists consistently fail to engage in questions of power and politics that sociologists, anthropologists, and science studies scholars engage.

Notes to Chapter 4

1. Interview, Oliver Ryder (July 20, 2005).

2. For a description of the practices of karyotyping by hand, and the transition to computer-based karyotyping, in human genetics, see Rapp (2000).

3. Interview, Oliver Ryder (July 20, 2005 and October 25, 2005); interview, Barbara Durrant (May 16, 2006); Informal conversation with someone who worked at CRES during that time.

4. See Stengers (2010) on the importance of slowing research down in order to conduct science in more democratic ways as part of her cosmopolitical proposal.

5. Interview, Oliver Ryder (July 20, 2005 and October 25, 2005).

6. Interview, Oliver Ryder (July 20, 2005).

7. See Rothfels (2002), and Hanson (2002).

8. See Baratay and Hardouin-Fugier (2002), and Hanson (2002).

9. See Hanson (2002), and Mitman (1999).

10. See Mitman (1996), and Rothfels (2002). For an example of this kind of critique of zoos, see Hancocks (2001).

11. See Hancocks (2001), and Jamieson (2002).

12. Rothfels (2002) shows how Hagenbeck's cages without bars were derived from his experience in exhibiting people who originated from regions remote to Europe. These exhibits were part of the complex connections between displaying and seeing "exotic" animals in zoos and human "freaks" in circuses, carnivals, and fairs. What generated support for these exhibits among both professional anthropologists and the general public was the illusion of personal freedom, despite the fact that these people were essentially treated as property, servants, and slaves. Hagenbeck brought this illusion of freedom developed in

human exhibits to the zoological park's display of animals for the purpose of entertainment and profit.

13. The efforts to breed bison in captivity at the Bronx Zoo can be seen as a precursor to this activity (Hanson 2002).

14. From the Minnesota Zoo website, http://www.mnzoo.org/conservation/conservation_atZootigerSSP.asp (Retrieved August 13, 2009).

15. See http://www.aza.org/species-survival-plan-program/ (Retrieved October 7, 2011). See also Braverman (2012a, 2012b) on Species Survival Plans.

16. Chrulew (2011) has noted that the facilitation of life through the biopolitics of SSPs does not mean that death is not also pursued, which is—I would add—consistent with Foucault's (2003) delineation of biopower. He notes that some zoos kill animals that are surplus to the SSP. I did not encounter culling in U.S. zoos, where such practices are not supported (Kaufman 2012).

17. See Benirschke (1986).

18. See also Haraway (2003a) on the linkages and fissures in biodiversity discourses and practices in Species Survival Plans when compared to dog breeding protocols.

19. These tissue samples had to be taken due to federal laws governing the movement of animals from the San Diego Zoo to the Wild Animal Park (Interview, Oliver Ryder, July 20, 2005).

20. Tour of CRES (July 20,2005). There was also a discussion about further elaborating such exchanges between field and zoo scientists at the 2006 meeting of the Feline Taxonomic Advisory Group. See also Benson (2010) on the ways in which wild animals are collected and released in the knowledge practices of field-based conservationists.

21. This is rather unfortunate because the banteng is probably much better suited to the conditions of the zoo than the polar bear, a species that many people do not believe should be held in captivity because they roam so widely (Clubb and Mason 2003).

22. Interview, Linda Penfold (April 17, 2006); interview, Sharon Joseph (October 18, 2005); interview, Oliver Ryder (October 25, 2005); interview, Philip Damiani (July 1, 2005).

23. Interview, Linda Penfold (April 17, 2006).

24. Interview, Sharon Joseph (October 18, 2005).

25. Interview, Linda Penfold (April 17, 2006).

26. For a philosophical discussion that develops the notion of scientific progress upon which such an argument is based, see Lakatos (1978). This linear vision of scientific knowledge has, however, been thoroughly problematized in science studies by emphasizing the path dependences of knowledge and technologies. See Latour (1999), and Hess (1997) for a discussion of how and why "progress" has been problematized in science studies. See Rheinberger (1997, 2010) for a critique of the notion that there is a linear relationship, and thus divide, between basic and applied science.

27. This sentiment is continuous with other areas of social life, including sociological perspectives on medical technologies and feminist perspectives on reproductive technologies. See Thompson (2005) on shifts in feminist perspectives on reproductive technologies as well as Timmermans (2000), and Timmermans and Marc Berg (2003) on shifting perspectives on medical technologies within sociology.

28. See also Bowker (2005) on biodiversity databases as space-saving technologies.

29. The centrality of space in this articulation of cloning helps to further explain why some Species Survival Plan managers were so angered by the cloned gaur and African wildcats. In these projects, scientists working in zoos or biotechnology companies decided what animals to produce as part of technology development. In doing so, they disregarded the work that Species Survival Plan managers do. In turn, the genetically redundant animals that resulted from these high-profile experiments were actually experienced as a problem for this group of zoo workers.

Notes to Chapter 5

1. Drawing on life science research regarding horizontal gene transfers, Helmreich (2003) has argued that sex is being replaced with transfer in contemporary biopolitics. This is true in the case of the transpositions of technology developers. However, it is not true in the case of the genetic values of plural technology developers. For a feminist science studies discussion of horizontal gene transfer, see also Hird (2009).

2. See Franklin (1997a, 2007b).

3. See Ritvo (1995), who has also emphasized that Bakewell was uncertain of the mechanisms through which traits were passed down through the generations, which were no better understood by Charles Darwin almost a century later when he developed his theory of evolution in the *Origin of Species* ([1859] 1996).

4. See also Orland (2003).

5. See Müller-Wille and Rheinberger (2007), Maienschien (1991, 2001, 2002, 2003), Ritvo (1995), and Franklin (2007b).

6. See Müller-Wille and Rheinberger (2007), and De Renzi (2007).

7. These developments only began to converge into "biology" during the middle of the nineteenth century (Müller-Wille and Rheinberger 2007: 9).

8. See also Strathern (1992a).

9. See Franklin (2007b).

10. This has been deemed a site of genetic essentialism (Best and Kellner 2002; Hopkins 1998; Nelkin and Lindee 1998; Nerlich, Clarke, and Dingwall 1999; Petersen 2002). However, even in agricultural contexts, genetic essentialism is not necessarily an appropriate critique. For example, one agricultural researcher told me how cloning was envisioned as a means to reproduce cattle with the highest-grade carcass score. At this point in time, carcass scores cannot be determined before death, making it impossible to selectively breed those select

cows who end up having the highest carcass score. Here, somatic cell nuclear transfer becomes a tool through which carcass score could be rationalized as a selected trait. That is, somatic cells could be taken from the slaughtered cow after death and after the assessment of a high carcass score. The somatic cell could then be used to create another individual of the same genetic composition. Resulting cloned bulls would serve as studs in selective breeding protocols, thereby making carcass score a selected trait. This raised an important question, however: is carcass score an inherited trait that is determined by one's genetic configuration? I asked this scientist what was known about the correlation between genomic inheritance and carcass score. He responded that his research group was collecting that data. They had estimated that the heritability of carcass traits was 60 percent. Some of the cloned animals that they produced and were monitoring had similar growth rates as the somatic cell donor. That said, he continued that they recognize that the environment was going to play a very important role in how the cloned animal "performs." As such, a significant portion of the research on somatic cell nuclear transfer in the agricultural industry has sought to understand the extent to which genetic inheritance contributes to the constitution of highly desired and deeply capitalized animal traits. Techniques like cloning gain their value by increasing the probability of reproducing animals whose carcass will grade out at the highest quality indicator.

11. An important exception to this would of course be the giant panda. China owns almost all giant pandas, and charges zoos a large sum to display these animals. Zoos are frequently willing to pay these fees to increase the number of visitors to the park, as the panda is probably the most charismatic of endangered species today. See Blacker (2012) for a commentary on the costs and politics of receiving two giant pandas from China in Edinburgh.

12. See Sunder Rajan (2006: 41) for a discussion of the links between the word value and ethics.

13. Bowker (2005) has discussed the move to molecularize memory in conservation practices. He argues that this is part of an attempt to "database the world" (Bowker 2005: 107) so as to remember the past. Frozen zoos clearly represent this type of endeavor, in that they seek to remember the genetic diversity of the present in cryopreserved form for future scientists to study. Bowker points out that every memory system must incorporate forgetting in order to make the practices of remembering feasible. Frozen zoos must forget the embodied animal in order to save genetic information from so many different species. However, Bowker believes that this kind of memory practice may forget that which matters most: processes of change. This is also what many conservationists are most concerned with—conserving conditions wherein evolution can continue to create biodiversity. This has been bound up in the changing terrain of environmentalism, which has moved from preserving static entities as they exist within a particular time (e.g., species) to preserving conditions through which a diversity of species can continue to evolve—and thereby change—into

the future (Takacs 1996: 67). Chrulew (2011) is similarly critical of the ways in which zoos privilege the genetic over the spatial, which he argues results in sickly animals. Taken together, these represent critiques of genetic essentialism in the park. What I want to add is that this genetic essentialism is "strategically" deployed in order for zoos to change themselves institutionally.

14. Worldwide it is estimated that half a billion people go to the zoo each year (Bertman 2004).

15. See also Brown (2003), Franklin (2003a, 2007), and Franklin and Lock (2003).

16. Interview, Linda Penfold (April 17, 2006).

17. For a more activist-oriented articulation of bio-piracy, see Shiva (1995). For empirical, case studies of how these extraction processes have been operationalized and resisted in practice see Anderson (2000), Hayden (2003), Soto Laveage (2009), Reardon (2005), and TallBear (2007).

18. I would like to thank Lynn Morgan for this statement, which she made at the workshop "The Anthropology of Potentiality: Exploring the Productivity of the Undefined and its Interplay with Notions of Humanness in New Medical Practices." The workshop was hosted by the Wenner-Gren Foundation at the Hotel Rosa dos Ventos, Teresopolos, Brazil, October 28–November 4, 2011.

19. Many species engage in relations that result in domestication and the presence of Homo sapiens is not required (Noske 1997; Anderson 1998a; O'Connor 1997). For example, Noske (1997: 3) points out that some species of ants "milk" aphids. Citing Donald R. Griffin, Noske (1997: 176) explains: "Certain species of ants feed on the sugary faeces exuded by aphids, a kind of plant-house which adheres to the plants. The ants gather and care for their 'domesticates' and sometimes even build shelters around them."

20. See Anderson (1998a, 1998b).

21. See Sutherland (2006) for a description of what it is like to work with zoo animals, which certainly highlights that these animals are not tame.

22. See Hoage, Roskell, and Mansour (1996), Baratay and Hardouin-Fugier (2002), Rothfels (2002), and Ritvo (1987, 1996).

23. Harris (1996: 440) has noted that "domestic" has long been used in archaeology, history, and anthropology in a static manner, denoting a "state of being" or a "kind" of animal. Meanwhile, biologists have understood domestication as a process. The contributors to *Where the Wild Things Are Now* (2007), edited by Cassidy and Mullin, use practice to link a process-based definition of domestication in biology with sociocultural studies.

24. See also Haraway (2008).

25. I would like to thank Rene Almeling and Jeff Ostergren for taking me and my partner to the new and old zoos in Los Angeles. At the time, zoo-goers could go inside the cages of the old zoo, including the animals' sleeping quarters that would normally be off limits. Strewn with litter and covered with graffiti, it nonetheless provides a fascinating window into the history of zoos.

26. See Rothfels (2002), and Mitman (1996).

Notes to Chapter 6

1. See Weale (2007) on the precautionary principle in environmental politics, which has been evident in discussions on both global warming (O'Riordan and Andres 1995; Westra 1997) and genetically modified foods (Jasanoff 2005). Inadequate scientific knowledge here serves as an argument to slow technology development so that risks can be assessed. Critics of cloning call on the precautionary principle when they argue that there is insufficient knowledge regarding embryonic development and the consequences of interspecies nuclear transfer. Weale (2007: 591) notes that the precautionary principle has been supported by policy statements, legislation, and international treaties and agreements. However, the principle has nonetheless been critiqued for being one-sided. One technology developer I interviewed made a similar criticism, noting that so much attention was focused on the risks of developing and using assisted reproductive technology with endangered animals while no one was asking about the risks of *not* developing and using these techniques.

2. See Wang, Swanson, Herrick et al. (2009).

3. See Holt, Pickard, and Prather (2004).

4. The Leverhulme Trust is one of the largest private research funders in the United Kingdom. Early career grants are competitive funds supporting postdoctoral researchers asking basic scientific questions.

5. In a review of the potential benefits and pitfalls of cloning endangered animals, Holt and his colleagues (2004) argued that somatic cell nuclear transfer could theoretically be used to conserve genetic diversity in small populations, in a manner that aligns with the logic embodied by the banteng. However, they argued that the great many eggs required to do nuclear transfer combined with the terribly low success rate of both nuclear transfer and embryo transfer greatly delimits this potential. Given this obstacle, Holt and his colleagues suggested that the best animals to clone are amongst species that release several eggs in a cycle and give birth to litters as opposed to singletons. According to these criteria, endangered mammals are not good candidates for cloning experimentation. Rather, Holt and his colleagues suggested that endangered rodents, amphibians, and fish are better species with which to develop this reproductive technology. The message of the review was that, if zoos are going to learn to clone, they should do so with species whose biology mitigates the technical problems associated with somatic cell nuclear transfer. In this context, Holt decided that amphibians were a particularly good species to clone because they produce large numbers of eggs externally, and so interspecies nuclear transfer may not be required for this taxa (Interview, May 20, 2010). As such, Holt used of "likelihood to succeed" as a criterion in deciding what species to clone, in a manner that aligns with technology developers. However, he looked across the range of different kinds of species to ask which had biological features that were most likely to offset the key problems associated with somatic cell nuclear

transfer. Technology developers instead focused on the sociotechnical aspects of species in making these decisions.

6. See also Holmes (1993) on the prominent role of frogs in experimental physiology.

7. See also Franklin (1999).

8. Stuart, Chanson, Cox et al. (2004: 1783) report that 32.5 percent of amphibian species are currently considered threatened by the IUCN Red List of Vulnerable, Endangered, or Critically Endangered Species. Meanwhile, 43.2 percent of amphibian species are declining.

9. See Wildt (2004) for an argument that reproductive scientists need to respond to conservation priorities in order to ensure that their research is timely and useful to the wider conservation community.

10. This endeavor has required significant research regarding how to cryopreserve frog cells. While somatic cells are normally taken from mammals for cryopreservation, these cells tend to be too contaminated with bacteria in frogs. Current research is indicating that the frog's eye, which grows cells that retain barriers protecting it from bacteria, is the best target cell for cryopreservation. Holt has, however, struggled to get funding for this research, which is more clearly driven by conservation needs. As such, Holt and Lloyd's work for the Amphibian Ark is often done voluntarily through support from the Zoological Society of London.

11. See Stuart, Chanson, Cox et al. (2004). It is important to note, however, that this kind of managed relocation is highly contested. See Minteer and Collins (2010), and Schwartz, Hellman, McLachlan et al. (2012).

12. Beebee and Grittiths (2005: 281) note that the successful reintroduction of species on the brink of extinction is an important way of demonstrating what originally caused the decline and that this problem has been identified and rectified. This is particularly important with amphibians because the reason why so many populations are declining is not well understood at this point in time.

13. See Howard, Marinari, and Wildt (2003) for a description of the role of reproductive science in the preservation of black-footed ferret.

14. See Macnaghten, Kearnes, and Wynne (2005), Rabinow and Bennett (2008), and Thompson (2013).

15. See also Wildt (2004: 288).

16. See also Clarke (1998).

17. It is important to emphasize that basic physiological knowledge can also come out of technology development. After all, Dolly the Sheep was an exercise in technology development that radically altered basic scientists' understandings of what occurs at a molecular level as cells developed (Franklin 1999, 2007b). My intention is not to reinstate a science/technology divide, but rather to draw out the consequences of different points of reference and emphasis. Prioritizing technology development versus basic, physiological knowledge production does

inform how projects are articulated, along with the criteria with which such projects are evaluated.

18. Changing "normal" developmental processes has been a key feature of experimental approaches in the life sciences. See Maienschien (2003), and Pauly (1987) for a discussion.

19. See Ankeny (2007) on the centrality of comparison in animal modeling. On the ways in which comparison is often made implicit through presumptions of conservation, see Logan (2001, 2002), and Rheinberger (2010: 6).

20. See also Ritvo (1997).

21. On the choice of model organisms, see also Burian (1993), Creager (2002), and Holmes (1993).

22. On animal models in the life sciences and biomedicine, see Ankeny (2007), Bolker (2009), Burian (1993), Bynum (1990), Clarke (1987, 2004), Creager (2002), Creager, Lunbeck, and Wise (2007), Clause (1993), Davies (2012), Hanson (2004), Haraway (1989, 1997, 2008), Kohler (1994), Lederman and Burian (1993), Logan (1999, 2001, 2002, 2005), Löwy (1992; 2000), (Löwy and Gaudilliere (1998), Rader (2004), Rheinberger (2010), and Thompson (2013).

23. Rader (2004) has provided the most thorough historical analysis of this process, focusing on the development of mice as research materials.

24. See also Holmes (1993).

25. See Clarke and Fujimura (1992a) on the right tools for scientific jobs. See Burian (1993), and Lederman and Burian (1993) for an extension of this concept to animal models.

26. See Takacs (1996).

27. Many corners of environmentalism argue that nonhuman life forms are intrinsically valuable and reject an approach to conservation that is rooted in the maintenance of resources for human needs. This perspective is particularly associated with deep ecology, which contends that humans are not intrinsically more valuable than nonhuman fauna and flora. For a discussion of intrinsic as opposed to instrumental values in environmentalism, see Benton (2007), Eckersley (1992), McKibben ([1989] 2003), Naess (1989), and Wapner (2010).

28. See Merchant (1980).

29. Indeed, the interspecies version of nuclear transfer pursued in zoos was the subject of sustained ethical scrutiny by the Human Fertilisation and Embryology Authority in the United Kingdom (2007). This legislative body decided that licenses could be awarded to institutions wanting to create human-animal "chybrid" embryos through interspecies nuclear transfer in the context of human embryonic stem cell research. However, no institution has sought such a license, in large part because the potential of interspecies nuclear transfer has been exceeded by the potential of induced pluripotent stem (iPS) cells (Emily Jackson, personal conversation, March 2011; James Porter, personal conversation, August 2009). For a social science discussion of chybrid embryo debates in the United Kingdom, see Brown (2009a), and Twine (2010). Hinterberger

(2011) is also initiating a new research project that looks at different national politics and policies on hybrid embryos.

30. I would like to thank Nikolas Rose for the path dependence metaphor in considering knowledge practices.

Notes to Chapter 7

1. Takacs (1996) notes that Aldo Leopold (1887–1948) helped to articulate this way of valuing diversity for scientific reasons, whether it be the diversity of organisms, landscapes, or cultures. Takacs states (1996: 13): "Science depicts the complexity of what Leopold refers to as 'the land organism'—today we might call it the ecosystem. . . . The policy prescription that follows is: save it all, for you know not what you do." This is closely linked to discourses that seek to preserve species and habitats on the basis that they could lead to economic development. Significantly, Takacs (1996: 55) points out that the fact that some species may generate economic value works to support the idea that all species and habitats should be preserved because one doesn't know what could be valuable or when.

2. See Takacs (1996).

3. See Keller (2000, 2010), Landecker (2010, 2011), and Powell and Dupré (2009).

4. For a discussion of complexity in relation to different kinds of reductionisms, see Law and Mol (2002), Nowotny (2005), Powell and Dupré (2009), and Wynne (2005).

5. See Anderson (2002), Nelkin and Lindee (1995), Nerlich, Dingwall, and Clarke (2002).

6. See Keller (2010). One of the more spectacular images in epigenetics is based on a study wherein mice gestating genetically identical offspring were given different diets during pregnancy. The offspring of the mouse given a normal diet were plump and yellow in color. The offspring of the mouse given a diet high in supplements was smaller and brown. See Morgan, Sutherland, Martin, and Whitelaw (1999) for the report. See Landecker (2010, 2011) for a discussion. What this means is that traits generally considered "genetic" are proving to be products of gene-environment interactions. Indeed, the first cloned cat—named CC for carbon copy—did not look like the somatic cell donor because cat coloration is not genetically determined.

7. See also Mitman (1999).

8. See Mitman (1999: 100), and Takacs (1996). Mitman (1999: 107) has pointed out that preserving individual species often intersected with the entertainment value of animals, whereas ecosystem approaches are situated in the links between the quality of land and the quality of life. While preserving species directs attention to modalities for managing animals, an ecosystem approach focuses on land management.

9. Landecker (2011: 178) notes a parallel phenomenon with nutritional epigenetics. It has long been known within epidemiology that poor health markers in early life are associated with poor health markers later in life. However, the

mechanisms or causes for such associations have been difficult to determine. Epigenetics is able to offer a molecular mechanism for these connections.

10. See Müller-Wille and Rheinberger (2007), and Keller (2010).

11. See Keller (2010).

12. See also Landecker (2011).

13. See also Shostak's (2004) research on the molecularization of human environments in the context of environmental health concerns, focusing specifically on the rise of toxicogenomics. She notes that the eclipse of space and place is also contested in this context.

14. See Hanson (2002).

15. See Rowell (1972) for a classic comparison of in situ and ex situ behavior among monkeys.

16. See Snyder, Derrickson, Beissinger et al. (1996) on the limitations of captive breeding for reintroduction.

17. See Benson (2010) for a discussion of the reintroduction of Keiko the killer whale, following the public outcry created by the film *Free Willy*. Benson (2010: 187) notes that a team member involved in the release noted that the whale was neither captive nor wild, but rather something in between. Benson (2010: 187) comments that most animals live in such a state today.

18. See also Wielebnowski, Ziegler, Wildt et al. (2002).

19. I am here drawing on Latour's (2004b) critique of social constructionism as a mode of critique and his corresponding development of "matters of concern."

20. For exemplars of this in the context of primate research, see Rowell (1972), and Strum and Fedigan (2000).

21. See also Friese (in press) on how different care practices are also being understood as constitutive of the laboratory within mechanistically based, life science research.

22. In making this argument, I am building on the long-standing use of the symmetry principle (Bloor 1991) in science studies. If the environment is constitutive in the wild, it must also be treated as constitutive in captivity.

23. I am here drawing on the BBC documentary series entitled *Unnatural Histories* (Nightingale and Murray 2011).

24. See Franklin (2002) for a theoretical elaboration of this idea that nature is everywhere.

25. See also Wapner (2010: 206) on the importance of ambiguity, the uncontrollable, and the unknown is preserving nature within the postnatural moment.

26. Vinciane Despret (2005) has offered the most thorough analysis of the importance of surprise in conducting responsible knowledge practices. In doing so, she has herself turned to the work of primatologist turned sheep watcher Thelma Rowell, who has been discussed throughout this chapter. Despret starts with the fact that, in order to answer questions about sheep behavior, Rowell introduced a twenty-third bowl of breakfast to the feeding schedule of her twenty-two sheep. Rowell's goal was to introduce an element of surprise into the

daily interactions of her sheep, which could elicit surprising findings regarding sheep behavior. Rowell's experiment was not based on control, but was instead meant to allow sheep the chance to do something interesting in relation to their social organization. Fostering surprise as opposed to control thus shaped the interactions between Rowell and her sheep within the experiment. However, surprise was not based on novelty per se. Rather, surprise provided a way to articulate responsible knowledge practices. Specifically, facilitating surprise meant that Rowell determined her questions by asking what the sheep might be concerned with. Rather than starting with what other scientists are interested in, Rowell started with the things that her sheep were interested in. Despret has argued that this could serve as a model for forging responsible knowledge practices across the social and life sciences, which start with the concerns of those being studied. Starting with the concerns of study participants has been central to participatory action research in sociological, public health, and medical research involving human subjects. Through the trope of surprise, Despret has been able to extend this approach to research involving nonhuman animals. Both are based on the researcher forgoing control in order to pursue more egalitarian relationships.

Notes to Conclusion

1. See Franklin (2003b) for an argument that it is not the parts that are crucial for social life and our studies of it, but rather the ways in which parts considered distinct are related.
2. See the Wellcome Trust's report on human cloning (1998), Williamson (1999), Eskridge and Stein (1998) for a bioethical discussion. The incredibly popular, fictional book *Never Let Me Go* (2005) by Ishiguro articulates this fear well.
3. See Brown (2003, 2009b), Brown, Rappert and Webster (2000), Van Lente (1993), and Wyatt (2004).
4. See Svendsen (2011).
5. For a critique of this discourse, see Haraway (2003a, 2003b, 2008).
6. See, for example, Irwin (2001), and Macnaghten and Urry (1998).
7. See Puig de la Bellacasa (2011) for an excellent discussion of this facet of Latour's argument.
8. See Lorimer and Driessen (in press) on the biopolitics of human relations with bovine. See Chrulew (2011) on the biopolitics of the park. See Sutherland (2006) for a popular discussion of the ways in which zoo animals are subjectified through stimulation exercises, and Haraway (2003, 2008) for a theoretical discussion.
9. Landecker (2007: 223) has, for example, noted that the ways human cells are cultured today has more to do with the ways that cells are cultured across species. The human-animal distinction is not nearly as important as the cell, tissue, organ distinction. In addition, the human-animal distinction becomes

increasingly difficult to sustain when considering the practices of cell culture, given that the culture medium sustaining cells is generally made up of animal parts.

10. I am here drawing on Nikolas Rose's (2007) argument that biopolitics is currently in shift due to five key changes, including: 1) a focus on the molecular level; 2) a focus on technologies that improve the stuff of life; 3) making persons responsible for their own health; 4) a shift in expertise from the state to the laboratory and its related appendages; and 5) the focus on commercialization and capitalization that undergirds each of the prior processes. For Rose, this is a distinctly human affair (Rabinow and Rose 2006). I would, however, contend that endangered animal cloning is without a doubt embroiled in at least four of these dynamics. It is difficult to say if endangered zoo animals have been made to feel responsible for their reproduction, per se.

BIBLIOGRAPHY

Adams, Vincanne, Michelle Murphy, and Adele E. Clarke. 2009. Anticipation: Technoscience, life, affect, temporality. *Subjectivity* 28:246–265.

Agamben, Giorgio. 1998. *Homo Sacer: Sovereign Power and Bare Life.* Translated by D. Heller-Roazen. Stanford: Stanford University Press.

Akrich, Madeline. 1992. The description of technical object. In *Shaping Technology/Building Society: Studies in Sociotechnical Change*, edited by W. E. Bijker and J. Law. Cambridge: MIT Press.

Anderson, Alison. 2002. In search of the Holy Grail: Media discourse and the new human genetics. *New Genetics and Society* 21 (3):327–337.

Anderson, Julie. 2001. Cloned ox dies; Scientists not fazed. *Omaha World Herald*, 12 January.

Anderson, Kay. 1998a. Animal domestication in geographic perspective. *Society & Animals* 6 (2):119–135.

———. 1998b. Animals, science, and spectacle in the city. In *Animal Geographies: Place, Politics, and Identity in the Nature-Culture Borderlands*, edited by J. Wolch and J. Emel. London: Verso.

Anderson, Warwick. 2000. The possession of kuru: Medical science and biocolonial exchange. *Comparative Studies in Society and History* 42 (713–743).

Ankeny, Rachel A. 2007. Wormy logic: Model organisms as case-based reasoning. In *Science without Laws: Model Systems, Cases, Exemplary Narratives*, edited by A. N. H. Creager, E. Lunbeck, and M. N. Wise. Durham: Duke University Press.

Associated Press. 2000. Scientists focus on Iowa cow carrying cloned ox fetus; 1 of 32 implants: If Bessie delivers, there's hope for saving endangered, even extinct, species. *Telegraphy Herald*, A1.

Baratay, Eric, and Elisabeth Hardouin-Fugier. 2002. *Zoo: A History of Zoological Gardens in the West.* Translated by O. Welsh. London: Reaktion Books.

Becker, Gay. 2000. *The Elusive Embryo: How Women and Men Approach New Reproductive Technologies.* Berkeley: University of California Press.

Beebee, Trevor J. C., and Richard A. Griffiths. 2005. The amphibian decline crisis: A watershed for conservation biology? *Biological Conservation* 125:271–285.

Benirschke, Kurt, ed. 1986. *Primates: The Road to Self-Sustaining Populations.* New York: Springer-Verlag.

Benjamin, Ruha. 2013. *People's Science: Bodies and Rights on the Stem Cell Frontier*. Stanford: Stanford University Press.

Benson, Etienne. 2010. *Wired Wilderness: Technologies of Tracking and the Making of Modern Wildlife*. Baltimore: Johns Hopkins University Press.

Benton, Ted. 2007. Deep ecology. In *The Sage Handbook of Environment and Society*, edited by J. Pretty, A. S. Ball, T. Benton, J. S. Guivant, D. R. Lee, D. Orr, M. J. Pfeffer, and H. Ward. London: Sage.

Berger, John. 2008. *Ways of Seeing*. London: Penguin.

Bertman, Brian. 2004. Misconceptions about zoos. *Biologist* 51 (4):199–206.

Best, Steven, and Douglas Kellner. 2002. Biotechnology, ethics and the politics of cloning. *Democracy & Nature* 8 (3):439–465.

Beyhan, Zeki, Amy E. Iager, and Jose B. Cibelli. 2007. Interspecies nuclear transfer: Implications for embryonic stem cell biology. *Cell Stem Cell* 1:502–512.

Biggins, Dean E., Astrid Vargas, Jerry L. Godbey, and Stanley H. Anderson. 1999. Influence of prerelease experience on reintroduced black-footed ferrets. *Biological Conservation* 89:121–129.

Bijker, Wiebe E., Thomas P. Hughes, and Trevor Pinch. 1987. *The Social Construction of Technological Systems: New Directions in the Sociology and History of Technology* Cambridge: MIT Press.

Birke, Lynda, Arnold Arluke, and Mike Michael. 2007. *The Sacrifice: How Scientific Experiments Transform Animals and People*. West Lafayette, Ind.: Purdue University Press.

Blacker, Terence. 2012. Pandas—the world's most political animal. Panda diplomacy has always been a squalid business. *The Independent*.

Bloor, David. 1991. *Knowledge and Social Imagery*. 2nd ed. Chicago: University of Chicago Press.

Bolker, Jessica A. 2009. Exemplary and surrogate models: Two modes of representation in biology. *Perspectives in Biology and Medicine* 52 (4):485–499.

Bolker, Jessica A., and Rudolf A. Raff. 1997. Beyond worms, flies and mice: It's time to widen the scope of developmental biology. *The Journal of NIH Research* 9:35–39.

Bolnick, Deborah A., Duana Fullwiley, Troy Duster, Richard S. Cooper, Joan H. Fujimura, Jonathan Kahn, Jay S. Kaufman, Jonathan Marks, Ann Morning, Alondra Nelson, Pilar Ossario, Jenny Reardon, Susan M. Reverby, and Kimberly TallBear. 2007. The science and business of genetic ancestry testing. *Science* 318:399–400.

Bostrom, Nick. 2004. The future of human evolution. In *Death and Anti-Death: Two Hundred Years after Kant, Fifty Years after Turing*, edited by C. Tandy. Palo Alto, Calif.: Ria University Press.

Bowker, Geoffrey C. 2005. *Memory Practices in the Sciences*. Cambridge: MIT Press.

Bowker, Geoffrey C., and Susan Leigh Star. 1999. *Sorting Things Out: Classification and Its Consequences*. Cambridge: MIT Press.

Brand, Stewart. 2009. *Whole Earth Discipline*. London: Atlantic Books.

Braun, Kathrin. 2007. Biopolitics and Temporality in Arendt and Foucault. *Time & Society* 16 (1):5–23.

Braverman, Irus. 2012a. *Zooland: The Institution of Captivity*. Palo Alto: Stanford University Press.

———. 2012b. Zooveillance: Controlling zoo animals to conserve. *Surveillance & Society* 10 (2):119–133.

Brown, Nik. 2003. Hope against hype: Accountability in biopasts, presents and futures. *Science Studies* 16 (2):3–21.

———. 2009a. Beasting the embryo: The metrics of humanness in the transspecies embryo debate. *BioSocieties* 4 (2–3):147–163.

———. 2009b. Shifting tenses: From regimes of truth to regimes of hope. *Configurations* 13 (3):331–355.

Brown, Nik, Brian Rappert, and Andrew Webster, eds. 2000. *Contested Futures: A Sociology of Prospective Techno-Science*. Aldershot: Ashgate.

Brubaker, Rogers, Mara Loveman, and Peter Stamatov. 2004. Ethnicity as cognition. *Theory and Society* 33 (1):31–64.

Burian, Richard M. 1993. How the choice of experimental organism matters: Epistemological reflections on an aspect of biological practice. *Journal of the History of Biology* 26 (2):351–368.

Butler, Judith. 1993. *Bodies That Matter: On the Discursive Limits of "Sex."* New York: Routledge.

Bynum, W. F. 1990. C'est une malade: Animal models and concepts in human disease. *Journal of the History of Medicine and Allied Sciences* 45 (3):397–413.

Callahan, Daniel. 1998. Cloning: Then and now. *Cambridge Quarterly of Healthcare Ethics* 7:141–144.

Callon, Michel. 1987. Society in the making: The study of technology as a tool for sociological analysis. In *The Social Construction of Technological Systems: New Directions in the Sociology and History of Technology*, edited by W. E. Bijker, T. P. Hughes, and T. Pinch. Cambridge: MIT Press.

———. [1986] 1999. Some elements of a sociology of translation: Domestication of the scallops and the fishermen of St. Brieuc Bay. In *The Science Studies Reader*, edited by M. Biaglio. New York: Routledge.

Canguilhem, Georges. 1978. *On the Normal and the Pathological*. Translated by C. R. Fawcett. Boston: Dr. Reidel.

Casper, Monica J. 1998. *The Making of the Unborn Patient: A Social Anatomy of Fetal Surgery*. New Brunswick: Rutgers University Press.

Casper, Monica J., and Lisa Jean Moore. 2009. *Missing Bodies: The Politics of Visibility*. New York: NYU Press.

Cassidy, Rebecca. 2002. *The Sport of Kings: Kinship, Class and Thoroughbred Breeding in Newmarket*. Cambridge: Cambridge University Press.

Cassidy, Rebecca, and Molly Mullin, eds. 2007. *Where the Wild Things Are Now: Domestication Reconsidered*. Oxford: Berg.

Charmaz, Kathy. 2000. Grounded theory: Objectivist and constructivist methods. In *Handbook of Qualitative Research*, edited by N. K. Denzin and Y. S. Lincoln. Thousand Oaks, Calif.: Sage.

Chen, Da-Yuan, Duan-Cheng Wen, Ya-Ping Zhang, Ging-Yuan Sun, Zhi-Ming Han, Zhong-Hua Liu, Peng Shi, Jin-Song Li, Jing-Gong Xiangyu, Li Lian, Zhao-Hui Kou, Yu-Qi Wu, Yu-Cun Chen, Peng-Yan Wang, and He-Min Zhang. 2002. Interspecies implantation of mitochondria fate of panda-rabbit cloned embryos. *Biology of Reproduction* 67 (2):637–642.

Chen, Mel. 2010. Animals without genitals: Race and transsubstantiation. *Women in Performance: A Journal of Feminist Theory* 20 (3):285–297.

Chrulew, Matthew. 2011. Managing love and death at the zoo: The biopolitics of endangered species preservation. *Australian Humanities Review* 50:137–157.

Claridge, M. F., H. A. Dawah, and M. R. Wilson. 1997. Practical approaches to species concepts for living organisms. In *Species: The Units of Biodiversity*, edited by M. F. Claridge, H. A. Dawah, and M. R. Wilson. London: Chapman & Hall.

Clark, Nigel. 1997. Panic ecology: Nature in the age of superconductivity. *Theory, Culture & Society* 14 (1):77–96.

———. 1999. Wild life: Ferality and the frontier with chaos. In *Quicksands: Foundational Histories in Australia and Aotearoa New Zealand*, edited by K. Neumann, N. Thomas, and H. Ericksen. Sydney: University of New South Wales Press.

Clarke, Adele E. 1987. Research materials and reproducive science in the United States, 1910–1940. In *Physiology in the American Context, 1850–1940*, edited by G. L. Geison. Bethesda, Md.: American Physiological Society.

———. 1991. Social worlds/arenas theory as organizational theory. In *Social Organization and Social Process: Essays in Honor of Anselm Strauss*, edited by D. Maines. Hawthorne, N.Y.: Aldine de Gruyter.

———. 1995. Modernity, postmodernity, & reproductive processes, ca. 1890–1990. Or "Mommy, where do cyborgs come from anyway?" In *The Cyborg Handbook*, edited by C. H. Gray. New York: Routledge.

———. 1998. *Disciplining Reproduction: Modernity, American Life Sciences, and the Problems of Sex*. Berkeley: University of California Press.

———. 2000. Maverick reproductive scientists and the production of contraceptives, 1915–2000+. In *Bodies of Technology: Women's Involvement with Reproductive Medicine*, edited by A. Rudinow Saetnan, N. Oudshoorn, and M. Kirejczyk. Columbus: Ohio State University Press.

———. 2003. Situational analyses: Grounded theory mapping after the postmodern turn. *Symbolic Interaction* 26 (4):553–576.

———. 2004. Reproductive physiological research at the department of embryology. In *100 Years of the Department of Embryology of the Carnegie Institution of Washington*, edited by J. Mainenschein and A. Garland. Cambridge: Cambridge University Press.

———. 2005. *Situational Analysis: Grounded Theory after the Postmodern Turn*. Thousand Oaks, Calif.: Sage.

Clarke, Adele E., and Carrie Friese. 2007. Grounded theorizing using situational analysis. In *Handbook of Grounded Theory*, edited by A. Bryant and K. Charmaz. London and Thousand Oaks: Sage.

Clarke, Adele E., and Joan H. Fujimura. 1992a. *The Right Tools for the Job: At Work in Twentieth-Century Life Sciences*. Princeton: Princeton University Press.

Clarke, Adele E., and Joan H. Fujimura. 1992b. What tools? Which jobs? Why right? In *The Right Tools for the Job: At Work in the Twentieth Century Life Sciences*, edited by A. E. Clarke and J. H. Fujimura. Princeton: Princeton University Press.

Clarke, Adele E., and Theresa Montini. 1993. The many faces of RU486: Tales of situated knowledges and technological contestations. *Science, Technology and Human Values* 18 (1):42–78.

Clarke, Adele E., Janet K. Shim, Laura Mamo, Jennifer Ruth Fosket, and Jennifer R. Fishman. 2003. Biomedicalization: Technoscientific transformations of health, illness, and U.S. biomedicine. *American Sociological Review* 68:161–194.

Clarke, Adele E., and Susan Leigh Star. 2003. Symbolic interactionist science, technology, information and biomedicine studies. In *Handbook of Symbolic Interaction*, edited by N. Herman and L. Reynolds. Walnut Creek, Calif.: AltaMira Press.

———. 2007. The social worlds framework: A theory/methods package. In *Handbook of Science & Technology Studies*, edited by by Edward J. Hackett, Olga Amsterdamska, Michael E. Lynch, and Judy Wajcman. Cambridge: MIT Press.

Clause, Bonnie Tocher. 1993. "The wistar rat as a right choice": Establishing mammalian standards and the ideal of a standardized mammal. *Journal of the History of Biology* 26 (2):329–350.

Clubb, R., and G. Mason. 2003. Animal welfare: Captivity effects on wide-ranging carnivores. *Nature* 425:473–474.

Colen, Shellee. 1995. "Like a mother to them": Stratified reproduction and West Indian childcare workers and employers in New York. In *Conceiving the New World Order: The Global Politics of Reproduction*, edited by F. D. Ginsburg and R. Rapp. Berkeley: University of California Press.

Creager, Angela N. H. 2002. *The Life of a Virus: Tobacco Mosaic Virus as an Experimental Model, 1930–1965*. Chicago: University of Chicago Press.

Creager, Angela N. H., Elizabeth Lunbeck, and M. Norton Wise, eds. 2007. *Science without Laws: Model Systems, Cases, Exemplary Narratives*. Durham: Duke University Press.

Cronon, William. 1997. *Uncommon Ground: Rethinking the Human Place in Nature*. New York: Norton.

D'Andrade, Roy. 1995. *The Development of Cognitive Anthropology*. Cambridge: Cambridge University Press.

Darwin, Charles. [1859] 1996. *The Origin of Species*. Oxford: Oxford University Press.

Davies, Gail. 2012. What is a humanized mouse? Remaking the species and spaces of translational medicine. *Body & Society* 18:126–155.

Davis-Floyd, Robbie. 1992. *Birth as an American Rite of Passage*. Berkeley: University of California Press.

De Renzi, Silvia. 2007. Resemblance, paternity, and imagination in early modern courts. In *Heredity Produced: At the Crossroads of Biology, Politics, and Culture, 1500–1870*, edited by S. Müller-Wille and H. J. Rheinberger. Cambridge: MIT Press.

Derry, Margaret E. 2003. *Bred for Perfection: Shorthorn Cattle, Collies, and Arabian Horses since 1800*. Baltimore: Johns Hopkins University Press.

Despret, Vinciane. 2004. The body we care for: Figures of anthropo-zoo-genesis. *Body & Society* 10 (2):111–134.

———. 2005. Sheep do have opinions. In *Making Things Public*, edited by B. Latour and P. Weibel. Cambridge: MIT Press.

———. 2008. The becomings of subjectivity in animal worlds. *Subjectivity* 23:123–139.

DiMaggio, Paul. 1997. Culture and cognition. *Annual Review of Sociology* 23:263–287.

Douglas, Mary. [1966] 2005. *Purity and Danger*. London: Routledge.

Durkheim, Emile. [1912] 1995. *The Elementary Forms of Religious Life*. Translated by K. E. Fields. New York: Free Press.

Durkheim, Emile, and Marcel Mauss. [1903] 1963. *Primitive Classification*. Translated by R. Needham. London: Cohen & West.

Dyson, Freeman. 2006. Make me a hipporoo: When children start to play with real genes, evolution as we know it will change forever, argues physicist and futurist (Transcript). *New Scientist*, April 25.

Eckersley, Robyn. 1992. *Environmentalism and Political Theory*. London: University College.

Edwards, Derek. 1991. Categories are for talking: On the cognitive and discursive bases of categorization. *Theory and Psychology* 1:515–542.

Edwards, Paul, and Christopher Lee. 2006. Infrastructuration: Technology Studies and/ as Social Theory. Paper read at Society for Social Studies of Science, at Vancouver, B.C.

Edwards, Jeanette. 1999. Why Dolly matters: Kinship, culture and cloning. *Ethnos* 64 (3):301–324.

Eggler, Bruce. 2011. Director of Audubon's endangered species research center is leaving. *The Times-Picayune*, February 19, 2011.

Epstein, Steven. 2007. *Inclusion: The Politics of Difference in Medical Research*. Chicago: University of Chicago Press.

———. 2008. Culture and science/technology: Rethinking knowledge, power, materiality and nature. *The Annals of the American Academy of Political and Social Science* 619:165–181.

Eskridge, William N., and Edward Stein. 1998. Queer clones. In *Clones and Cloning: Facts and Fantasies about Human Cloning*, edited by M. Nussbaum and C. Sustein. New York: Norton.

Evans, John H. 2010. *Contested Reproduction: Genetic Technologies, Religion, and Public Debate*. Chicago: University of Chicago Press.

Firestone, Shulasmith. 1970. *The Dialectic of Sex. The Case for a Feminist Revolution*. New York: Bantam.

Foster, Laura. 2010. Re-inventing hoodia: Patent law, epistemic citizenship, and the making of difference in South Africa. Ph.D. dissertation, University of California, Los Angeles.

———. 2011. Situating feminism, patent law, and the public domain. *Columbia Journal of Gender and Law* 20 (1):261–347.

———. 2012. Patents, biopolitics, and feminisms: Locating patent law struggles over breast cancer genes and the hoodia plant. *International Journal of Cultural Property* 19 (3): 371–400.

Fothergill. 2006. *Planet Earth.* U.K.

Foucault, Michel. 1978. *The History of Sexuality. Volume 1: An Introduction.* Translated by R. Hurley. New York: Vintage Books.

———. 2003. *Society Must Be Defended. Lectures at the College de France 1975–1976.* Translated by D. Macey. Edited by M. Bertani, A. Fontana, F. Ewald, and A. I. Davison. New York: Picador.

———. [1966] 1970. *The Order of Things: An Archaeology of the Human Sciences.* New York: Vintage Books.

Fox News/Opinion Dynamics Poll. *Do you think it is acceptable to use cloning (http://www.pollingreport.com/science.htm)* 2002 [cited June 4, 2007].

Franklin, Adrian. 2002. *Nature and Social Theory.* London: Sage.

Franklin, Sarah. 1997a. Dolly: A new form of transgenic breedwealth. *Environmental Values* 6:427–437.

———. 1997b. *Embodied Progress: A Cultural Account of Assisted Conception.* London: Routledge.

———. 1999. Review essay: What we know and what we don't about cloning and society. *New Genetics and Society* 18:111–120.

———. 2001. Biologization revisited: Kinship theory in the context of the new biologies. In *Relative Values: Reconfiguring Kinship Studies*, edited by S. Franklin and S. McKinnon. Durham: Duke University Press.

———. 2003a. Ethical biocapital: New strategies of cell culture. In *Remaking Life & Death: Toward an Anthropology of the Biosciences*, edited by S. Franklin and M. Lock. Santa Fe: School of American Research Press and James Currey.

———. 2003b. Re-thinking nature-culture. *Anthropological Theory* 3 (1):65–85.

———. 2003c. Kinship, genes, and cloning: Life after Dolly. In *Genetic Nature/Culture: Anthropology and Science beyond the Two-Culture Divide*, edited by A. H. Goodman, D. Heath and M. S. Lindee. Berkeley: University of California Press.

———. 2006. The cyborg embryo: Our path to transbiology. *Theory, Culture & Society* 23 (7–8):167–187.

———. 2007a. Analogic return. In *American Anthropological Association Annual Meetings.* San Francisco, Calif.

———. 2007b. *Dolly Mixtures: The Remaking of Genealogy.* Durham: Duke University Press.

———. 2010. Transbiology: A feminist cultural account of being after IVF. In *Critical Conceptions: Technology, Justice, and the Global Reproductive Market Conference.* Barnard Center for Research on Women, New York.

Franklin, Sarah, and Margaret Lock, eds. 2003. *Remaking Life & Death: Toward an Anthropology of the Biosciences.* Santa Fe: School of American Research Press.

Franklin, Sarah, Celia Lury, and Jackie Stacey. 2000. *Global Nature, Global Culture.* London: Sage.

Franklin, Sarah, and Helena Ragone, eds. 1998. *Reproducing Reproduction: Kinship, Power, and Technological Innovation.* Philadelphia: University of Pennsylvania Press.

Franklin, Sarah, and Celia Roberts. 2006. *Born and Made: An Ethnography of Preimplantation Genetic Diagnosis.* Princeton: Princeton University Press.

Friese, Carrie. 2009. Models of cloning, models for the zoo: Rethinking the sociological significance of cloned animals. *BioSocieties* 4 (4):367–390.

———. In press. Realizing potential in translational medicine: The uncanny emergence of care as science. *Current Anthropology.*

Friese, Carrie, and Adele E. Clarke. 2012. Transposing bodies of knowledge and technique: Animal models at work in the reproductive sciences. *Social Studies of Science* 42 (1):31–52.

Fuglie, Keith O., Clare A. Narrod, and Catherine Neumeyer. 2000. Public and private investments in animal research. In *Public-Private Collaboration in Agricultural Research: New Institutional Arrangements and Economic Implications,* edited by K. O. Fuglie and D. E. Schimmelpfennig. Ames: Iowa State University Press.

Fujimura, Joan H. 1992. Crafting science: Standardized packages, boundary objects, and "translation." In *Science as Practice and Culture,* edited by A. Pickering. Chicago: University of Chicago Press.

———. 1996. *Crafting Science: A Socio-History of the Quest for the Genetics of Cancer.* Cambridge: Harvard University Press.

Fukuyama, Francis. 2002. *Our Posthuman Future: Consequencs of the Biotechnology Revolution.* New York: Picador.

Fuller, Steve. 2011. *Humanity 2.0: What It Means to Be Human: Past, Present and Future.* Basingstoke: Palgrave Macmillan.

Ganchoff, Chris. 2004. Regenerating movements: Embryonic stem cells and the politics of potentiality. *Sociology of Health & Illness* 26 (6):757–774.

Gascon, Claude, James P. Collins, Robin D. Moores, Don R. Church, Jeanne E. McKay, and Joseph R. Mendelson III. 2005. Amphibian conservation action plan. Gland, Switzerland and Cambridge: IUCN/SSC Amphibian Conservation Summit.

Gaskell, George. 2000. Agricultural biotechnology and public attitudes in the European Union. *AgBioForum* 3 (2 & 3):87–96.

Geesink, Ingrid, Barbara Prainsack, and Sarah Franklin. 2008. Stem cell stories 1998–2008. *Science as Culture* 17 (1):1–11.

Gieryn, Thomas. 1995. Boundaries of science. In *Handbook of Science and Technology Studies,* edited by S. Jasanoff, G. Markle, J. Petersen, and T. Pinch. Thousand Oaks, Calif.: Sage.

Ginsburg, Faye D., and Rayna Rapp, eds. 1995. *Conceiving the New World Order: The Global Politics of Reproduction.* Berkeley: University of California Press.

Gomez, Martha C., Jill A. Jenkins, Angelica Giraldo, Rebecca F. Harris, Amy King, Betsy L. Dresser, and Charles Earle Pope. 2003. Nuclear transfer of synchronized African wild cat somatic cells into enucleated domestic cat oocytes. *Biology and Reproduction* 69 (3):1032–1041.

Gomez, Martha C., C. Earle Pope, Angelica Giraldo, Leslie A. Lyons, Rebecca F. Harris, Amy L. King, Alex Cole, Robert A. Godke, and Betsy L. Dresser. 2004. Birth of African wildcat cloned kittens born from domestic cats. *Cloning and Stem Cells* 6:247–258.

Gomez, Martha C., Charles Earle Pope, Robert H. Kutner, David M. Ricks, Leslie A. Lyons, Mark Ruhe, Cherie Dumas, Justine Yons, Monica Lopez, Betsy L. Dresser, and Jakob Reiser. 2008. Nuclear transfer of sand cat cells into enucleated domestic cat oocytes is affected by cryopreservation of donor cells. *Cloning and Stem Cells* 10 (4):469–483.

Gottweiss, Herbert, Brian Salter, and Catherine Waldby. 2009. *The Global Politics of Human Embryonic Stem Cell Science*. Hampshire: Palgrave Macmillan.

Gunton, Mike. 2009. *Life*. U.K.

Habermas, Jurgen. 2003. *The Future of Human Nature*. Cambridge: Polity.

Hacking, Ian. 1999. *The Social Construction of What?* Cambridge: Harvard University Press.

———. 2006. Making up people. *London Review of Books* 28 (16):23–26.

Hancocks, David. 2001. *A Different Nature: The Paradoxical World of Zoos and Their Uncertain Future*. Berkeley: University of California Press.

Hanson, Elizabeth. 2002. *Animal Attractions: Nature on Display in American Zoos*. Princeton: Princeton University Press.

———. 2004. How Rhesus monkeys became laboratory animals. In *Centennial History of the Carnegie Institution of Washington: The Department of Embryology*, edited by J. Maienschein, M. Glitz and G. E. Allen. Cambridge: Cambridge University Press.

Haraway, Donna J. 1989. *Primate Visions: Gender, Race, and Nature in the World of Modern Science*. New York: Routledge.

———. 1991. *Simians, Cyborgs, and Women: The Reinvention of Nature*. New York: Routledge.

———. 1997. *Modest_Witness@Second_Millenium.FemaleMan©_Meets_OncoMouse™: Feminism and Technoscience*. New York: Routledge.

———. 2003a. Cloning mutts, saving tigers: Ethical emergents in technocultural dog worlds. In *Remaking Life & Death: Toward an Anthropology of the Biosciences*, edited by S. Franklin and M. Lock. Santa Fe and Oxford: School of American Research Press and James Currey.

———. 2003b. *The Companion Species Manifesto: Dogs, People, and Significant Otherness*. Chicago: Prickly Paradigm Press.

———. 2008. *When Species Meet*. Minneapolis: University of Minnesota Press.

Harrington, Anne, Nikolas Rose, and Ilina Singh. 2006. Editors' Introduction. *BioSocieties* 1 (1):1–5.

Harris, D. R. 1996. Domesticatory relationships of people, plants and animals. In *Redefining Nature*, edited by R. Ellen. Oxford: Berg.

Hartouni, Valerie. 1993. *Brave New World* in the discourses of reproductive and genetic technologies. In *In the Nature of Things: Language, Politics, and the Environment*, edited by J. Bennett and W. Chaloupka. Minneapolis: University of Minnesota Press.

———. 1997. *Cultural Conceptions: On Reproductive Technologies & the Remaking of Life.* Minneapolis: University of Minnesota Press.

Hayden, Corinne. 2003. *When Nature Goes Public: The Making and Unmaking of Bioprospecting in Mexico.* Princeton: Princeton University Press.

Heatherington, Terry. 2008. Cloning the wild mouflon. *Anthropology Today* 24 (1):9–14.

Helmreich, Stefan. 2003. Trees and seas of information: Alien kinship and the biopolitics of gene transfer in marine biology and biotechnology. *American Ethnologist* 30 (3):340–358.

Hess, David J. 1997. *Science Studies: An Advanced Introduction.* New York: NYU Press.

Hill, J. R., and I. Dobrinski. 2006. Male germ cell transplantation in livestock. *Reproduction, Fertility and Development* 18 (1–2):13–18.

Hinterberger, Amy. 2011. Human-animal chimeras. In *Discussion Paper for the Constitutional Foundations of Bioethics: Cross-National Comparisons.* Cambridge: Harvard University.

Hird, Myra J. 2009. *The Origins of Sociable Life: Evolution after Science Studies.* Basingstoke: Palgrave Macmillan.

Hoage, R. J., Anne Roskell, and Jane Mansour. 1996. Menageries and zoos to 1900. In *New Worlds, New Animals: From Menagerie to Zoological Park in the Nineteenth Century*, edited by R. J. Hoage and W. A. Deiss. Baltimore: Johns Hopkins University Press.

Hochadel, Oliver. 2011. Watching exoctic animals next door: "Scientific" observations at the zoo (ca. 1870-1910). *Science in Context* 24 (2):183–214.

Holmes, Frederic L. 1993. The old martyr of science: The frog in experimental physiology. *Journal of the History of Biology* 26 (2):6–13.

Holt, William V., Amanda R. Pickard, and Randall S. Prather. 2004. Wildlife conservation and reproductive cloning. *Society for Reproduction and Fertility* 127:317–324.

Hopkins, Patrick D. 1998. Bad copies: How popular media represent cloning as an ethical problem. *Hastings Center Report* 28 (2):6–13.

Howard, J., P. Marinari, and D. Wildt. 2003. Black-footed ferret: Model for assisted reproductive technologies contributing to in situ conservation. In *Reproductive Science and Integrated Conservation*, edited by W. V. Holt, A. R. Pickard, J. Rodger, and D. D. Wildt. Cambridge: Cambridge University Press.

Hradecky, P., J. Stover, and G. G. Stott. 1988. Histology of a heifer placentome after interspecies transfer of a gaur embryo. *Theriogenology* 30 (3):593–604.

Human Fertilization & Embryology Authority. 2007. Hybrids and chimeras. London.

Inhorn, Marcia C. 2003. Global infertility and the globalization of new reproductive technologies: Illustrations from Egypt. *Social Science & Medicine* 56 (9):1837–1851.

Inhorn, Marcia C., and Daphna Birenbaum-Carmeli. 2008. Assisted reproductive technologies and culture change. *Annual Review of Anthropology* 37:177–196.

Irwin, Alan. 2001. *Sociology and the Environment.* Cambridge: Polity.

Ishiguro, Kazuo. 2005. *Never Let Me Go.* London: Faber and Faber.

Jackson, Emily. 2001. *Regulating Reproduction: Law, Technology, and Autonomy.* Oxford: Hart Publishing.

———. 2009. *Medical Law: Text, Cases, and Materials*. Oxford: Oxford University Press.

Jamieson, Dale. 2002. Against zoos. In *Morality's Progress: Essays on Humans, Other Animals, and the Rest of Nature*, edited by D. Jamieson. Oxford: Clarendon Press.

Janssen, D. L., M. L. Edwards, J. A. Koster, R. P. Lanza, and O. A. Ryder. 2003. Postnatal management of chyptorchid banteng calves cloned by nuclear transfer utilizing Frozen fibroblast cultures and enucleated cow ova (abstract). *Reproduction, Fertility and Development* 16 (2):224–224.

Jasanoff, Sheila. 2004. *States of Knowledge: The Co-Production of Science and Social Order*. London: Routledge.

———. 2005. *Designs on Nature: Science and Democracy in Europe and the United States*. Princeton: Princeton University Press.

Jasanoff, Sheila, G. Markle, J. Petersen, and T. Pinch. 2004. Ordering knowledge, ordering society. In *States of Knowledge: The Co-Production of Science and Social Order*, edited by S. Jasanoff. London: Routledge.

Jewell, Elizabeth J., and Frank Abate, eds. 2001. *The New Oxford American Dictionary*. Oxford: Oxford University Press.

Jule, Kristen R., Lisa A. Leaver, and Stephen E. G. Lea. 2008. The effects of captive experience on reintroduction suvival in carnivores: A review and analysis. *Biological Conservation* 141:355–363.

Kaufman, Leslie. 2012. Zoos' bitter choice: To save some species, letting others die. *New York Times*.

Keller, Evelyn Fox. 1995. *Refiguring Life: Metaphors of Twentieth-Century Biology*. New York: Columbia University Press.

———. 2000. *The Century of the Gene*. Cambridge: Harvard University Press.

———. 2010. *The Mirage of a Space between Nature and Nurture*. Durham: Duke University Press.

Klotzko, Arlene Judity, ed. 2001. *The Cloning Sourcebook*. Oxford: Oxford University Press.

Kohler, R.E. 1994. *Lords of the Fly: Drosophila Genetics and the Experimental Life*. Chicago: University of Chicago Press.

Krogh, August. 1929. The progress of physiology. *Science* 70:200–204.

Lakatos, Imre. 1978. *The Methodology of Scientific Research Programmes*. Cambridge: Cambridge University Press.

Lakoff, George, and Mark Johnson. 1980. *Metaphors We Live By*. Chicago: University of Chicago Press.

Landecker, Hannah. 2007. *Culturing Life: How Cells Became Technologies*. Cambridge: Harvard University Press.

———. 2010. Food as exposure: Nutritional epigenetics and the molecular politics of eating. *Center for the Study of Women Update*, http://escholarship.org/uc/item/6j50v781.

———. 2011. Food as exposure: Nurtitional epigenetics and the new metabolism. *BioSocieties* 6 (2):167–194.

Lanza, Robert P., Jose B. Cibelli, Francisca Diaz, Carlos T. Moraes, Peter W. Farin, Charlotte E. Farin, Carolyn J. Hammer, Michael D. West, and Philip Damiani. 2000. Cloning of an endangered species using interspecies nuclear transfer. *Cloning* 2 (2):79–90.

Lanza, Robert P., Betsy L. Dresser, and Philip Damiani. 2000. Cloning Noah's ark: Biotechnology might offer the best way to keep some endangered species from disappearing from the planet. *Scientific American* November:85–89.

Laslett, Barbara, and Johanna Brenner. 1989. Gender and social reproduction: Historical perspectives. *Annual Review of Sociology* 15:381–404.

Latour, Bruno. 1987. *Science in Action: How to Follow Scientists and Engineers through Society*. Cambridge: Harvard University Press.

———. 1988. *The Pasteurization of France*. Cambridge: Harvard University Press.

———. 1993. *We Have Never Been Modern*. Cambridge: Harvard University Press.

———. 1999. *Pandora's Hope: Essays on the Reality of Science Studies*. Cambridge: Harvard University Press.

———. 2004a. *Politics of Nature: How to Bring the Sciences into Democracy*. Cambridge: Harvard University Press.

———. 2004b. Why has critique run out of steam? From matters of fact to matters of concern. *Critical Inquiry* 30:225-248.

———. 2005. *Reassembling the Social: An Introduction to Actor-Network-Theory*. Oxford: Oxford University Press.

Latour, Bruno, and Peter Weibel, eds. 2005. *Making Things Public: Atmospheres of Democracy*. Cambridge: MIT Press.

Law, John. 1999. After ANT: Complexity, naming and topology. In *Actor Network Theory and After*, edited by J. Law and J. Hassard. Maiden, Mass.: Blackwell.

Law, John, and Annemarie Mol, eds. 2002. *Complexities: Social Studies of Knowledge Practices*. Durham: Duke University Press.

Leach, Helen M. 2003. Human domestication reconsidered. *Current Anthropology* 44 (3):349-368.

———. 2007. Selection and the unforeseen consequences of domestication. In *Where the Wild Things Are Now: Domestication Reconsidered*, edited by R. Cassidy and M. Mullin. Oxford: Berg.

Lederman, M., and R. M. Burian. 1993. The right organism for the job: Introduction. *Journal of the History of Biology* 26 (2):235–238.

Levine, Donald A. 2002. Hybridization and extinction: In protecting rare species, conservations should consider the dangers of interbreeding, which compound the more well-known threats to wildlife. *American Scientist* 90 (3):254–258.

Lien, Marrianne. 2007. Domestication "Downunder": Atlantic salmon farming in Tasmania. In *Where the Wild Things are Now: Domestication Reconsidered*, edited by R. Cassidy and M. Mullins. Oxford: Berg.

Liu, Jennifer An-Hwa. 2008. Biomedtech nation: Taiwan, ethics, stem cells and other biologicals. Anthropology, University of California, San Francisco with University of California, Berkeley, Berkley and San Francisco.

Lock, Margaret. 2005. Eclipse of the gene and the return of divination. *Current Anthropology* 46 (Supplement):S47–S70.

Logan, Cheryl A. 1999. The altered rationale behind the choice of a standard animal in experimental psychology. Henry H. Donaldson, Adolf Meyer, and "the albino rat." *History of Psychology* 2 (1):3–4.

———. 2001. "[A]re Norway rats . . . things?"* Diversity versus generality in the use of albino rats in experiments on development and sexuality. *Journal of the History of Biology* 34 (2):287–314.

———. 2002. Before there were standards: The role of test animals in the production empirical generality in physiology. *History of the Journal of Biology* 35 (2):329–363.

———. 2005. The legacy of Adolf Meyer's comparative approach: Worcester rats and the strange birth of the animal model. *Integrative Physiological and Behavioral Science* 40 (4):169–181.

Loi, Pasqualino, Grazyna Ptak, Barbara Barboni, Josef Fulka Jr., Pietro Cappai, and Michael Clinton. 2001. Genetic rescue of an endangered mammal by cross-species nuclear transfer using post-mortem somatic cells. *Nature Biotechnology* 19:962–964.

Lorimer, Jamie. 2012. Multinatural geographies for the Anthropocene. *Progress in Human Geography* 36 (5):593–612.

Lorimer, Jamie, and Clemens Driessen. In press. Bovine biopolitics and the promise of monsters in the rewilding of Heck cattle. *Geoforum*.

Löwy, Ilana. 1992. From guinea pigs to man: The development of Haffkine's anticholera vaccine. *The Journal of the History of Medicine and Allied Science* 47:270–309.

———. 2000. The experimental body. In *Medicine in the Twentieth Century*, edited by J. Pickstone and R. Cooter. London: Routledge.

Löwy, Ilana, and John Paul Gaudilliere. 1998. Disciplining cancer: Mice and the practice of genetic purity. In *The Invisible Industrialist: Manufacturers and the Production of Scientific Knowledge*, edited by I. Löwy and J. P. Gaudilliere. London: Palgrave Macmillan.

Lynch, Michael. 1989. Sacrifice and the transformation of the animal body into a scientific object: Laboratory culture and ritual practice in the neurosciences. *Social Studies of Science* 18:265–289.

Lynch, Michael, and Kathleen Jordan. 2000. Patents, promotions, and protocols: Mapping and claiming scientific territory. *Mind, Culture, and Activity* 7 (1–2):124–146.

MacKenzie, Donald, and Judy Wajcman, eds. 1999. *The Social Shaping of Technology*. 2nd ed. Buckingham, U.K.: Open University Press.

Macnaghten, Phil, Matthew B. Kearnes, and Brian Wynne. 2005. Nanotechnology, governance, and public deliberation: What role for the social sciences. *Science Communication* 27 (2):268–291.

Macnaghten, Phil, and John Urry. 1998. *Contested Natures*. London: Sage.

Maienschein, Jane. 1991. The origins of *Entwicklungsmechanik*. In *A Conceptual History of Modern Embryology*, edited by S. F. Gilbert. Baltimore: Johns Hopkins University Press.

———. 2001. On cloning: Advocating history of biology in the public interest. *Journal of the History of Biology* 34:423–432.

———. 2002. What's in a name: Embryos, clones, and stem cells. *The American Journal of Bioethics* 2 (1):12–19.

———. 2003. *Whose View of Life? Embryos, Cloning, and Stem Cells*. Cambridge: Harvard University Press.

Mamo, Laura. 2007. *Queering Reproduction: Achieving Pregnancy in the Age of Technoscience*. Durham: Duke University Press.

Martin, Emily. 1987. *The Woman in the Body: A Cultural Analysis of Reproduction*. Boston: Beacon Press.

———. Forthcoming. *Reproduction*. In *Critical Terms for Gender Studies*, edited by C. R. Stimpson and G. Herdt. Chicago: University of Chicago Press.

Martin, Paul, Nik Brown, and Andrew Turner. 2008. Capitalizing hope: The commercial development of umbilical cord blood stem cell banking. *New Genetics and Society* 27(2): 128–143.

Marx, Karl. [1867] 1978. Capital, Volume One. In *The Marx-Engels Reader*, edited by R. C. Tucker. New York: Norton.

McKibben, Bill. [1989] 2003. *The End of Nature: Humanity, Climate Change and the Natural World*. London: Bloomsbury.

McPhee, M. Elsbeth. 2003. Generations in captivity increases behavioral variance: Considerations for captive breeding and reintroduction programs. *Biological Conservation* 115:71–77.

Mead, George Herbert. 1970. Self as social object. In *Social Psychology through Symbolic Interaction*, edited by G. Stone and H. Faberman. Waltham, Mass.: Xeros Publications.

Merchant, Carolyn. 1980. *The Death of Nature: Women, Ecology and the Scientific Revolution*. New York: Harper One.

Meyer, J., and E. Hunt. *International Species Information System website* 2008 [cited July 2009].

Milligan, Stuart R., William V. Holt, and Rhiannon Lloyd. 2009. Impacts of climate change and environmental factors on reproduction and development in wildlife. *Philosophical Transactions of the Royal Society* 364:3313–3319.

Minteer, Ben A., and James P. Collins. 2010. Move it or lose it? The ecological ethics of relocating species under climate change. *Ecological Applications* 20 (7):1801–1804.

Mitman, Gregg. 1996. When nature is the zoo: Vision and power in the art and science of natural history. *Osiris* 11:117–143.

———. 1999. *Reel Nature: America's Romance with Wildlife on Film*. Cambridge: Harvard University Press.

Morgan, Hugh D., Heidi G. E. Sutherland, David I. K. Martin, and Emma Whitelaw. 1999. Epigenetic inheritance at the agoui locus in the mouse. *Nature Genetics* 23:314–318.

Müller-Wille, Staffan, and Hans-Jörg Rheinberger, eds. 2007. *Heredity Produced: At the Crossroads of Biology, Politics, and Culture, 1500–1870*. Cambridge: MIT Press.

Naess, Arne. 1989. *Ecology, Community and Lifestyle*. Translated by D. Rothenberg. Cambridge: Cambridge University Press.

Nelkin, Dorothy. 1995. *Selling Science: How the Press Covers Science and Technology,* revised edition. New York: W. H. Freeman.

Nelkin, Dorothy, and M. Susan Lindee. 1995. *The DNA Mystique: The Gene as a Cultural Icon.* New York: W. H. Freeman.

———. 1998. Cloning in the popular imagination. *Cambridge Quarterly of Healthcare Ethics* 7:145–149.

Nelson, Alondra. 2008. Bioscience: Genetic genealogy testing and the pursuit of African ancestry. *Social Studies of Science* 38 (5):759–783.

Nerlich, Brigitte, David D. Clarke, and Robert Dingwall. 1999. The influence of popular cultural imagery on public attitudes towards cloning. *Sociological Research Online* 4 (3):http://www.socresonline.org.uk/socresonline/4/3nerlich.html.

Nerlich, Brigitte, Robert Dingwall, and David D. Clarke. 2002. The book of life: How the completion of the Human Genome Project was revealed to the public. *Health: An Interdisciplinary Journal for the Social Study of Health, Illness and Medicine* 6 (4):445–469.

Newman, Stewart. 2010. The transhumanism bubble. *Capitalism Nature Socialism* 21 (2):29–42.

Nightingale, Neil, and Andrew Murray. 2011. *Unnatural Histories.* United Kingdom: BBC Four.

Noske, Barbara. 1997. *Beyond Boundaries: Humans and Animals.* Montreal: Black Rose Books.

Nowotny, Helga. 2005. The increase of complexity and its reduction: Emergent interfaces between the natural sciences, humanities and social sciences. *Theory, Culture & Society* 22:15–31.

O'Connor, T. 1997. Working at relationships: Another look at animal domestication. *Antiquity* 71:149–156.

O'Riordan, Timothy, and Jordan Andres. 1995. The precautionary principle in contemporary enviornmental politics. *Enviornmental Values* 4 (3):191–212.

Oelschlaeger, Max. 1991. *The Idea of Wilderness: From Prehistory to the Age of Ecology.* New Haven: Yale University Press.

Oh, H. J., M. K. Kim, G. Jang, H. J. Kim, S. G. Hong, J. E. Park, K. Park, C. Park, S. H. Sohn, D.Y. Kim, N. S. Shin, and B. C. Lee. 2008. Cloning endangered gray wolves (Canis lupus) from somatic cells collected postmortem. *Theriogenology* 70 (4):638–647.

Orland, Barbara. 2003. Turbo-cows: Producing a competitive animal in the nineteenth and early twentieth centuries. In *Industrializing Organisms: Introducing Evolutionary History,* edited by S. R. Schrepfer and P. Scranton. New York: Routledge.

Oxford English Dictionary. 2009. Articulation. In *Oxford English Dictionary.* Oxford: Oxford University Press.

Pauly, Philip J. 1987. *Controlling Life: Jacques Loeb and the Engineering Ideal in Biology.* Oxford: Oxford University Press.

Petersen, Alan. 2002. Replicating our bodies, losing our selves: News media portrayals of human cloning in the wake of Dolly. *Body and Society* 8 (4):71–90.

Pickering, Andrew. 1995. *The Mangle of Practice: Time, Agency, & Science*. Chicago: University of Chicago Press.

Powell, Alexander, and John Dupré. 2009. From molecules to systems: The importance of looking both ways. *Studies in History and Philosophy of Biological and Biomedical Sciences* 40:54–64.

Prainsack, Barbara. 2006. Negotiating life: The regulation of human cloning and embryonic stem cell research in Israel. *Social Studies of Science* 36 (2):173–205.

Priest, Susanna Hornig. 2000. US public opinion divided over biotechnology? *Nature Biotechnology* 18:989–942.

———. 2001. Cloning: A study in news production. *Public Understanding of Science* 10:59–69.

Puig de la Bellacasa, Maria. 2011. Matters of care in technoscience: Assembling neglected things. *Social Studies of Science* 41 (1):85–106.

Rabinow, Paul. 1996. Artificiality and enlightenment: From sociobiology to biosociality. In *Essays on the Anthropology of Reason*. Princeton: Princeton University Press.

———. 1999. *French DNA: Trouble in Purgatory*. Chicago: University of Chicago Press.

———. 2000. Epochs, presents, events. In *Living and Working with the New Medical Technologies: Intersections of Inquiry*, edited by M. Lock, A. Young, and A. Cambrosio. Cambridge: Cambridge University Press.

Rabinow, Paul, and Gaymon Bennett. 2008. From bioethics to human practices, or assembling contemporary equipment. In *Tactical Biopolitics: Art, Activism, and Technoscience*, edited by B. da Costa and P. Kavita. Cambridge: MIT Press.

Rabinow, Paul, and Nikolas Rose. 2006. Biopower today. *BioSocieties* 1:195–217.

Rader, Karen. 2004. *Making Mice: Standardizing Animals for American Biomedical Research, 1900–1955*. Princeton: Princeton University Press.

Rapp, Rayna. 2000. *Testing Women, Testing the Fetus: The Social Impact of Amniocentesis in America*. New York: Routledge.

Reardon, Jenny. 2001. The Human Genome Diversity Project: A case study in coproduction. *Social Studies of Science* 31:357–388.

———. 2005. *Race to the Finish: Identity and Governance in an Age of Genomics*. Princeton: Princeton University Press.

———. 2007. Democratic mishaps: The problem of democratization in a time of biopolitics. *BioSocieties* 2:239–356.

Rheinberger, Hans-Jörg. 1997. *Toward a History of Epistemic Things: Synthesizing Proteins in the Test Tube*. Stanford: Stanford University Press.

———. 2000. Beyond nature and culture: Modes of reasoning in the age of molecular biology and medicine. In *Living and Working with the New Medical Technologies: Intersections of Inquiry*, edited by M. Lock, Allan Young and Alberto Cambosio. Cambridge: Cambridge University Press.

———. 2010. *An Epistemology of the Concrete: Twentieth-Century Histories of Life*. Durham: Duke University Press.

Ritvo, Harriet. 1987. *The Animal Estate: The English and Other Creatures in the Victorian Age*. Cambridge: Harvard University Press.

————. 1995. Possessing mother nature: Genetic capital in eighteenth-century Britain. In *Early Modern Conceptions of Property*, edited by J. Brewer and S. Staves. London: Routledge.

————. 1996. The order of nature: Constructing the collections of Victorian zoos. In *New Worlds, New Animals: From Menagerie to Zoological Park in the Nineteenth Century*, edited by R. J. Hoage and W. A. Deiss. Baltimore: Johns Hopkins University Press.

————. 1997. *The Platypus and the Mermaid, and Other Figments of the Classifying Imagination*. Cambridge: Harvard University Press.

Roberts, Dorothy. 1997. *Killing the Black Body: Race, Reproduction and the Meaning of Liberty*. New York: Pantheon Books.

Robinson, Michael H. 1996. Introduction. In *New Worlds, New Animals: From Menagerie to Zoological Park in the Nineteenth Century*, edited by R. J. Hoage and W. A. Deiss. Baltimore: Johns Hopkins University Press.

Rose, Nikolas. 2001. The politics of life itself. *Theory, Culture & Society* 18 (6):1–30.

————. 2007. *The Politics of Life Itself: Biomedicine, Power, and Subjectivity in the Twenty-First Century*. Princeton: Princeton University Press.

Rothfels, Nigel. 2002. *Savages and Beasts: The Birth of the Modern Zoo*. Baltimore: Johns Hopkins University Press.

Rowell, Thelma. 1972. *Social Behaviour of Monkeys*. Edited by B. M. Foss, *Penguin Science of Behaviour*. Middlesex: Penguin Books.

Ryder, Oliver A., and Kurt Benirschke. 1997. The potential use of "cloning" in the conservation effort. *Zoo Biology* 16:295–300.

Saetnan, Ann Rudinow, Nelly Oudshoorn, and Marta Kirejczyk, eds. 2000. *Bodies of Technology: Women's Involvement with Reproductive Medicine*. Columbus: Ohio State University Press.

Savulescu, Julian. 2010. Human liberation: Removing biological and psychological barriers to freedom. *Monash Bioethics Review* 29 (1):1–18.

Savulescu, Julian, and Nick Bostrom, eds. 2009. *Human Enhancement*. Oxford: Oxford University Press.

Schwartz, Mark W., Jessica J. Hellmann, Jason M. McLachlan, Dov F. Sax, Justin O. Borevitz, Jean Brennan, Alejandro E. Camacho, Gerardo Ceballos, Jamie R. Clark, Holly Doremus, Regan Early, Julie R. Etterson, Dwight Fielder, Jacquelyn L. Gill, Patrick Gonzalez, Nancy Green, Lee Hannah, Dale W. Jamieson, Debra Javeline, Ben A. Minteer, Jay Odenbaugh, Stephen Polasky, David M. Richardson, Terry L. Root, Hugh D. Safford, Osvaldo Sala, Stephen H. Schneider, Andrew R. Thompson, John W. Williams, Mark Vellend, Pati Vitt, and Sandra Zellmer. 2012. Managed relocation: Integrating the scientific, regulatory, and ethical challenges. *BioScience* 62 (8):732–743.

Shapin, Steven, and Simon Schaffer. 1985. *Leviathan and the Air-Pump: Hobbes, Boyle and the Experimental Life*. Princeton: Princeton University Press.

Shim, Janet. 2005. Constructing "race" across the science-lay divide: Racial formation in the epidemiology and experience of cardiovascular disease. *Social Studies of Science* 35 (3):405–436.

Shiva, V. 1995. Biotechnological development and the conservation of biodiversity. In *Biopolitics: A Feminist and Ecological Reader on Biotechnology*, edited by V. Shiva and I. Moser. Andhra Pradesh, India: Orient Longman.

Shostak, Sara. 2004. Environmental justice and genomics: Acting on the futures of environmental health. *Science as Culture* 13 (4):539–562.

Shukin, Nicole. 2009. *Animal Capital: Rendering Life in Biopolitical Times*. Minneapolis: University of Minnesota Press.

Snyder, Noel F. R., Scott R. Derrickson, Steven R. Beissinger, James W. Wiley, Thomas B. Smith, William D. Toone, and Brian Miller. 1996. Limitations of captive breeding in endangered species recovery. *Conservation Biology* 10 (2):228–248.

Soto Laveaga, Gabriela. 2009. *Jungle Laboratories: Mexican Peasants, National Projects, and the Making of the Pill*. Durham: Duke University Press.

Star, Susan Leigh. 1991. Power, technologies and the phenomenology of conventions: On being allergic to onions. In *A Sociology of Monsters? Essays on Power, Technology and Domination*, edited by J. Law. London: Routledge.

Star, Susan Leigh, and Karen Ruhleder. 1996. Steps toward an ecology of infrastructure: Design and access for large information spaces. *Information Systems Research* 7 (1):111–134.

Star, Susan Leigh, and Anselm Strauss. 1999. Layers of silence, arenas of voice: The ecology of visible and invisible work. *Computer Supported Cooperative Work* 8:9–30.

Starr, Paul. 1992. Social categories and claims in the liberal state. *Social Research* 59 (2):263–295.

Stengers, Isabelle. 2010. *Cosmopolitics I*. Translated by R. Bononno. Minneapolis: University of Minnesota Press.

Stoler, Ann Laura. 1991. Carnal knowledge and imperial power: Gender, race and morality in colonial Asia. In *Gender at the Crossroads of Knowledge: Feminist Anthropology in the Postmodern Era*, edited by M. di Leonardo. Berkeley: University of California Press.

Stover, J., J. Evans, and E. P. Dolensek. 1981. Interspecies embryo transfer from the gaur to domestic Holstein. In *Proc. Annual AAZV Meeting*. Seattle, Wash.

Strathern, Marilyn. 1992a. *After Nature: English Kinship in the Late Twentieth Century*. Cambridge: Cambridge University Press.

———. 1992b. *Reproducing the Future: Anthropology, Kinship and the New Reproductive Technologies*. Manchester: Manchester University Press.

Strauss, Anselm. 1988. The articulation of project work: An organizational process. *Sociological Quarterly* 29:163–178.

Strauss, Anselm, and Juliet Corbin. 1998. *Basics of Qualitative Research: Techniques and Procedures for Developing Grounded Theory*. Thousand Oaks, Calif.: Sage.

Strum, Shirley C., and Linda Marie Fedigan, eds. 2000. *Primate Encounters: Models of Science, Gender, and Society*. Chicago: University of Chicago Press.

Stuart, Simon N., Janice S. Chanson, Neil A. Cox, Bruce E. Young, Ana S. L. Rodrigues, Debra L. Fischman, and Robert W. Waller. 2004. Status and trends of amphibian declines and extinctions worldwide. *Science* 306:1783–1786.

Sunder Rajan, Kaushik. 2003. Genomic capital: Public cultures and market logics of corporate biotechnology. *Science as Culture* 12 (1):87–121.

———. 2006. *Biocapital: The Constitution of Postgenomic Life.* Durham: Duke University Press.

Sutherland, Amy. 2006. *Kicked, Bitten, and Scratched: Life and Lessons at the World's Premier School for Exotic Animal Trainers.* New York: Viking Books.

Suzuki, Yuka. 2007. Putting the lions out at night: Domestication and the taming of the wild. In *Where the Wild Things Are Now: Domestication Reconsidered,* edited by R. Cassidy and M. Mullin. Oxford: Berg.

Svendsen, Mette N. 2011. Articulating potentiality: Notes on the delineation of the blank figure in human embryonic stem cell research. *Cultural Anthropology* 26 (3):414–437.

Takacs, David. 1996. *The Idea of Biodiversity: Philosophies of Paradise.* Baltimore: Johns Hopkins University Press.

TallBear, Kim. 2007. Narratives of race and indigeneity in the Genographic Project. *Journal of Law, Medicine & Ethics* (Fall):412–424.

Thacker, Eugene. 2003. Data made flesh: Biotechnology and the discourse of the posthuman. *Cultural Critique* 53:72–97.

Thompson [Cussins], Charis. 1996. Ontological choreography: Agency through objectification in infertility clinics. *Social Studies of Science* 26 (3):575–610.

———. 1998. Producing reproduction: Techniques of normalization and naturalization in infertility clinics. In *Reproducing Reproduction: Kinship, Power, and Technological Innovation,* edited by S. Franklin and H. Ragone. Philadelphia: University of Pennsylvania Press.

———. 1999. Confessions of a bioterrorist: Subject position and reproductive technologies. In *Playing Dolly: Technocultural Formations, Fantasies, and Fictions of Assisted Reproduction,* edited by A. E. Kaplan and S. Squier. New Brunswick: Rutgers University Press.

Thompson, Charis. 2001. Strategic naturalizing: Kinship in an infertility clinic. In *Relative Values: Reconfiguring Kinship Studies,* edited by S. Franklin and S. McKinnon. Durham: Duke University Press.

———. 2002a. Ranchers, scientists, and grass-roots development in the United States and Kenya. *Environmental Values* 11:303–326.

———. 2002b. When elephants stand for competing philosophies of nature: Amboseli National Park, Kenya. In *Complexities: Social Studies of Knowledge Practices,* edited by J. Law and A. Mol. Durham: Duke University Press.

———. 2005. *Making Parents: The Ontological Choreography of Reproductive Technologies.* Cambridge: MIT Press.

———. 2013. *Good Science: The Ethical Choreography of Stem Cell Research.*

Timmermans, Stefan. 2000. Technology and medical practice. In *Handbook of Medical Sociology,* edited by C. Bird, P. Conrad and A. M. Fremont. Upper Saddle River, N.J.: Prentice Hall.

Timmermans, Stefan, and Marc Berg. 2003. The practice of medical technology. *Sociology of Health and Illness* 25:97–114.

Titmuss, Richard. [1971] 1997. *The Gift Relationship: From Human Blood to Social Policy.* London: London School of Economics.

Turner, Stephanie S. 2002. Jurassic park technology in the bioinformatics economy: How cloning narratives negotiate the telos of DNA. *American Literature* 74 (4):887–909.

———. 2008. Open-ended stories: Extinction narratives in genome time. *Literature and Medicine* 26 (1):55–82.

Twine, Richard. 2010. Genomic natures read through posthumanisms. *The Sociological Review* 58 (s1):175–195.

U.S. Fish & Wildlife. 1993. Species accounts: Florida panther, Felis concolor coryi (Bangs). In *Endangered and Threatened Species of the Southeastern United States (Red Book) FWS Region 4.* http://www.fsw.gve/endangered/i/a/saao5.html.

Van Lente, Harro. 1993. *Promising Technology: The Dynamics of Expectations in Technological Developments.* Enschede: University of Twente.

Veltre, Thomas. 1996. Menageries, metaphors, and meanings. In *New Worlds, New Animals: From Menagerie to Zoological Park in the Nineteenth Century,* edited by R. J. Hoage and W. A. Deiss. Baltimore: Johns Hopkins University Press.

Wainwright, Steven P., Clare Williams, Mike Michael, Bobbie Farsides, and Alan Cribb. 2006. From bench to bedside? Biomedical scientists' expectations of stem cell science as a future therapy for diabetes. *Social Science & Medicine* 63:2052–2064.

Waldby, Catherine. 2000. *The Visible Human Project: Informatic Bodies and Posthuman Medicine.* New York: Routledge.

———. 2002. Stem cells, tissue cultures and the production of biovalue. *Health: An Interdisciplinary Journal for the Social Study of Health, Illness and Medicine* 6 (3):305–323.

Waldby, Catherine, and Robert Mitchell. 2006. *Tissue Economies: Blood, Organs, and Cell Lines in Late Capitalism.* Durham: Duke University Press.

Wang, C., W. F. Swanson, J. R. Herrick, K. Lee, and Z. Machaty. 2009. Analysis of cat oocyte activation methods for the generation of feline disease models by nuclear transfer. *Reproductive Biology and Endocrinology* 7:148.

Wapner, Paul. 2010. *Living through the End of Nature: The Future of American Environmentalism* Cambridge: MIT Press.

Weale, Albert. 2007. The precautionary principle in environmental policies. In *The Sage Handbook of Environment and Society,* edited by J. Pretty, A. S. Ball, T. Benton, J. S. Guivant, D. R. Lee, D. Orr, M. J. Pfeffer, and H. Ward. London: Sage.

Wellcome Trust. 1998. *Public Perspectives on Human Cloning.* London: The Wellcome Trust.

Western, David. 2007. The delicate balance: Environment, economics, development. *Issues in Science and Technology* 16 (3):http://www.issues.org/16.3/western.htm.

Western, David, Shirley Strum, and R. Michael Wright. 1994. *Natural Connections: Perspectives in Community-Based Conservation.* Washington, D.C.: Island Press.

Westra, Laura. 1997. Post-normal science, the precautionary principle and the ethics of integrity. *Foundations of Science* 2:237–262.

Wielebnowski, Nadja C., Karen Ziegler, David D. Wildt, John Lukas, and Janine L. Brown. 2002. Impact of social management on reproductive, adrenal and behavioral activity in the cheetah. *Animal Conservation* 5:291–301.

Wildt, David E. 2004. Making wildlife research more meaningful by prioritizing science, linking disciplines, and building capacity. In *Experimental Approaches to Conservation Biology*, edited by M. S. Gordon and S. M. Bartol. Berkeley: University of California Press.

Wilkie, Tom, and Elizabeth Graham. 1998. Power without responsibility: Media portrayals of Dolly and science. *Cambridge Quarterly of Healthcare Ethics* 7:150–159.

Williams, Harrison A. 1973. *Endangered Species Act*. United States Fish and Wildlife Service and National Oceanic and Atmospheric Administration. Washington, D.C., USA.

Williamson, Robert. 1999. Human reproductive cloning is unethical because it undermines autonomy: Commentary on Savulescu. *Journal of Medical Ethics* 25:96–97.

Wilmot, Sarah. 2007. Between the farm and the clinic: Agriculture and reproductive technology in the twentieth century. *Studies in History and Philosophy of Biological and Biomedical Sciences* 38 (2):303–315.

Wilmut, Ian. 1999. The age of biological control, eugenics, and human rights. In *The Genetic Revolution and Human Rights*, edited by J. Burley. Oxford: Oxford University Press.

Wilson, E. O. 1984. *Biophilia*. Cambridge: Harvard University Press.

Winner, Landon. 1980. Do artefacts have politics? *Daedalus* 109:121–136.

World Association of Zoos and Aquariums. *Banteng*. http://www.waza.org/en/zoo/visit-the-zoo/cattle-1254385523/bos-javanicus [cited July 2012].

Wyatt, Sally. 2004. Danger! Metaphors at work in economics, geophysiology, and the internet. *Science, Technology and Human Values* 29 (2):242–261.

———. 2008. Technological determinism is dead; long live technological determinism. In *The Handbook of Science and Technology Studies*, edited by E. J. Hackett, O. Amerdamska, M. Lynch, and J. Wajcman. Cambridge: MIT Press.

Wynne, Brian. 2005. Reflexing complexity: Post-genomic knowledge and reductionist returns in public science. *Theory, Culture & Society* 22:67–94.

Yanagisako, Sylvia, and Carol Delaney, eds. 1995. *Naturalizing Power: Essays in Feminist Cultural Analysis*. New York: Routledge.

Zerubavel, Eviatar. 1991. *The Fine Line: Making Distinctions in Everyday Life*. New York: Free Press.

———. 1996. Lumping and splitting: Notes on social classification. *Sociological Forum* 11 (3):421–433.

———. 1997. *Social Mindscapes: An Invitation to Cognitive Sociology*. Cambridge: Harvard University Press.

ABOUT THE AUTHOR

Carrie Friese is Lecturer in Sociology at the London School of Economics and Political Science.